ADVANCED AVIONICS
HANDBOOK

ADVANCED AVIONICS HANDBOOK

FAA-H-8083-6

Federal Aviation Administration

SKYHORSE PUBLISHING

Skyhorse Publishing books may be purchased in bulk at special discounts for sales promotion, corporate gifts, fund-raising, or educational purposes. Special editions can also be created to specifications. For details, contact the Special Sales Department, Skyhorse Publishing, 307 West 36th Street, 11th Floor, New York, NY 10018 or info@skyhorsepublishing.com.

Skyhorse® and Skyhorse Publishing® are registered trademarks of Skyhorse Publishing, Inc.®, a Delaware corporation.

www.skyhorsepublishing.com

10 9 8 7 6 5 4 3 2 1

Library of Congress Cataloging-in-Publication Data

Advanced avionics handbook : FAA-H-8083-6 / Federal Aviation Administration.
    p. cm.
 Includes bibliographical references.
 ISBN 978-1-61608-533-9 (alk. paper)
 1. Avionics--Handbooks, manuals, etc. I. United States. Federal Aviation Administration.
 TL695.A24 2011
 629.135--dc23
                          2011038387

ISBN 978-1-61608-533-9

Printed in China

Preface

The *Advanced Avionics Handbook* is a new publication designed to provide general aviation users with comprehensive information on advanced avionics equipment available in technically advanced aircraft. This handbook introduces the pilot to flight operations in aircraft with the latest integrated "glass cockpit" advanced avionics systems.

Since the requirements can be updated and the regulations can change, the Federal Aviation Administration (FAA) recommends that you contact your local Flight Standards District Office (FSDO), where FAA personnel can assist you with questions regarding advanced avionics equipment flight training and/or advanced avionics equipment questions about your aircraft.

This publication is available free of charge for download, in PDF format, from the FAA Regulatory Support Division (AFS-600) on the FAA website at *www.faa.gov*.

The *Advanced Avionics Handbook* may also be purchased from:
> Superintendent of Documents
> United States Government Printing Office
> Washington, DC 20402-9325
> http://bookstore.gpo.gov

This handbook is published by and comments should be sent in email form to:
> afs630comments@faa.gov

Acknowledgments

The FAA wishes to acknowledge the following aviation manufacturers and companies that provided images used in this handbook:

Avidyne Corporation

Cirrus Design, Inc.

Garmin Ltd.

Rockwell Collins, Inc.

S-Tec Corporation

The FAA would also like to extend its appreciation to the General Aviation Manufacturers Association (GAMA) for its assistance and input in the preparation of this handbook.

Table of Contents

Chapter 1

Introduction to Advanced Avionics

Introduction

This handbook is designed as a technical reference for pilots who operate aircraft with advanced avionics systems. Whether flying a conventional aircraft that features a global positioning system (GPS) navigation receiver or a new aircraft with the latest integrated "glass cockpit" advanced avionics system, you should find this handbook helpful in getting started. The arrival of new technology to general aviation aircraft has generated noticeable changes in three areas: information, automation, and options.

Pilots now have an unprecedented amount of information available at their fingertips. Electronic flight instruments use innovative techniques to determine aircraft attitude, speed, and altitude, presenting a wealth of information in one or more integrated presentations. A suite of cockpit information systems provides pilots with data about aircraft position, planned route, engine health and performance, as well as surrounding weather, traffic, and terrain.

Advanced avionics systems can automatically perform many tasks that pilots and navigators previously did by hand. For example, an area navigation (RNAV) or flight management system (FMS) unit accepts a list of points that define a flight route, and automatically performs most of the course, distance, time, and fuel calculations. Once en route, the FMS or RNAV unit can continually track the position of the aircraft with respect to the flight route, and display the course, time, and distance remaining to each point along the planned route. An autopilot is capable of automatically steering the aircraft along the route that has been entered in the FMS or RNAV system. Advanced avionics perform many functions and replace the navigator and pilot in most procedures. However, with the possibility of failure in any given system, the pilot must be able to perform the necessary functions in the event of an equipment failure. Pilot ability to perform in the event of equipment failure(s) means remaining current and proficient in accomplishing the manual tasks, maintaining control of the aircraft manually (referring only to standby or backup instrumentation), and adhering to the air traffic control (ATC) clearance received or requested. Pilots of modern advanced avionics aircraft must learn and practice backup procedures to maintain their skills and knowledge. Risk management principles require the flight crew to always have a backup or alternative plan, and/or escape route. Advanced avionics aircraft relieve pilots of much of the minute-to-minute tedium of everyday flights, but demand much more initial and recurrent training to retain the skills and knowledge necessary to respond adequately to failures and emergencies.

The FMS or RNAV unit and autopilot offer the pilot a variety of methods of aircraft operation. Pilots can perform the navigational tasks themselves and manually control the aircraft, or choose to automate both of these tasks and assume a managerial role as the systems perform their duties. Similarly, information systems now available in the cockpit provide many options for obtaining data relevant to the flight.

Advanced avionics systems present three important learning challenges as you develop proficiency:

1. How to operate advanced avionics systems

2. Which advanced avionics systems to use and when

3. How advanced avionics systems affect the pilot and the way the pilot flies

How To Operate Advanced Avionics Systems

The first challenge is to acquire the "how-to" knowledge needed to operate advanced avionics systems. This handbook describes the purpose of each kind of system, overviews the basic procedures required to use it, explains some of the

logic the system uses to perform its function, and discusses each system's general limitations. It is important to note that this handbook is not intended as a guide for any one manufacturer's equipment. Rather, the aim is to describe the basic principles and concepts that underlie the internal logic and processes and the use of each type of advanced avionics system. These principles and concepts are illustrated with a range of equipment by different manufacturers. It is very important that the pilot obtain the manufacturer's guide for each system to be operated, as only those materials contain the many details and nuances of those particular systems. Many systems allow multiple methods of accomplishing a task, such as programming or route selection. A proficient pilot tries all methods, and chooses the method that works best for that pilot for the specific situation, environment, and equipment. Not all aircraft are equipped or connected identically for the navigation system installed. In many instances, two aircraft with identical navigation units are wired differently. Obvious differences include slaved versus non-slaved electronic horizontal situation indicators (EHSIs) or primary flight display (PFD) units. Optional equipment is not always purchased and installed. The pilot should always check the equipment list to verify what is actually installed in that specific aircraft. It is also essential for pilots using this handbook to be familiar with, and apply, the pertinent parts of the regulations and the Aeronautical Information Manual (AIM).

Advanced avionics equipment, especially navigation equipment, is subject to internal and external failure. You must always be ready to perform manually the equipment functions which are normally accomplished automatically, and should always have a backup plan with the skills, knowledge, and training to ensure the flight has a safe ending.

Which Advanced Avionics Systems To Use and When

The second challenge is learning to manage the many information and automation resources now available to you in the cockpit. Specifically, you must learn how to choose which advanced cockpit systems to use, and when. There are no definitive rules. In fact, you will learn how different features of advanced cockpit avionics systems fall in and out of usefulness depending on the situation. Becoming proficient with advanced avionics means learning to use the right tool for the right job at the right time. In many systems, there are multiple methods of accomplishing the same function. The competent pilot learns all of these methods and chooses the method that works best for the specific situation, environment, and equipment. This handbook will help you get started in learning this important skill.

How Advanced Avionics Systems Affect the Pilot

The third challenge is learning how advanced avionics systems affect the pilot. The additional information provided by advanced avionics systems can affect the way you make decisions, and the ability to automate pilot tasks can place you in the role of system supervisor or manager. These ideas are presented throughout the handbook using a series of sidebars illustrating some of the issues that arise when pilots work with advanced avionics systems. This series is not a complete list; rather, its purpose is to convey an attitude and a manner of thinking that will help you continue to learn.

The Learning series provides tips that can help expedite mastery of advanced avionics. You will learn why taking the time to understand how advanced systems work is a better learning strategy than simply memorizing the button-pushing procedures required to use each system. The importance of committing to an ongoing learning process will be explained. Because of the limits of human understanding, together with the quirks present in computerized electronic systems of any kind, you will learn to expect, and be prepared to cope with, surprises in advanced systems. Avionics equipment frequently receives software and database updates, so you must continually learn system functions, capabilities, and limitations.

The Awareness series presents examples of how advanced avionics systems can enhance pilot awareness of the aircraft systems, position, and surroundings. You will also learn how (and why) the same systems can sometimes decrease awareness. Many studies have demonstrated a natural tendency for pilots to sometimes drift out of the loop when placed in the passive role of supervising an FMS/RNAV and autopilot. You will learn that one way to avoid this pitfall is to make smart choices about when to use an automated system, and when to assume manual control of the flight; how cockpit information systems can be used to keep you in touch with the progress of the flight when automated systems are used; and how some advanced cockpit systems can be set to operate in different modes, with each mode exhibiting a different behavior. Keeping track of which modes are currently in use and predicting the future behavior of the systems is another awareness skill that you must develop to operate these aircraft safely.

The Risk series provides insights on how advanced avionics systems can help you manage the risk faced in everyday flight situations. Information systems offer the immediate advantage of providing a more complete picture of any situation, allowing you to make better informed decisions about potential hazards, such as terrain and weather. Studies have shown that these same systems can sometimes have a negative effect on pilot risk-taking behavior. You will learn about situations in which having more information can tempt you to take more risk than you might be willing to accept without the information. This series will help you use advanced information systems to increase safety, not risk. As much as advanced information systems have improved the information stream to the cockpit, the inherent limitations of the information sources and timeliness are still present; the systems are not infallible.

When advanced avionics systems were first introduced, it was hoped that those new systems would eliminate pilot error. Experience has shown that while advanced avionics systems do help reduce many types of errors, they have also created new kinds of errors. This handbook takes a practical approach to pilot error by providing two kinds of assistance in the form of two series: Common Errors and Catching Errors. The Common Errors series describes errors commonly made by pilots using advanced avionics systems. These errors have been identified in research studies in which pilots and flight instructors participated. The Catching Errors series illustrates how you can use the automation and information resources available in the advanced cockpit to catch and correct errors when you make them.

The Maintaining Proficiency series focuses on pilot skills that are used less often in advanced avionics. It offers reminders for getting regular practice with all of the skills you need to maintain in your piloting repertoire.

Chapter Summary

This introductory chapter provided a broad perspective into the advanced avionics now found in many aircraft. This new equipment relieves the pilot of some tedious tasks while adding new ones and the requirement for more preflight study to learn the advanced capabilities and how to use the features. The pilot now has more and sometimes better means of fixing position, but has to contend with greater data loss when equipment breaks. It is important to maintain proficiency with the standby instruments and be proficient with the emergency tasks associated with the advanced avionics. Since these are electrical devices, the electrical generation and backup systems on the aircraft are even more important than ever.

Advanced avionics generally incorporate displays allowing pictures of the flight route as well as basic flight instrument data. While this can be most helpful to you, it can also lead you into areas where the pilot has no recourse, if any circumstances such as weather or equipment operation changes for the worse. You should never fly further into marginal conditions with advanced avionics than you would fly with conventional instruments. Advanced avionics do not enable an aircraft and pilot to break the laws of physics.

Advanced avionics were designed to increase safety as well as utility of the aircraft. Safety is enhanced by enabling better situational awareness. Safety can be increased by providing more information for you in an easier to interpret presentation.

Safety of flight can be hampered if you are not aware of what data the presentation is displaying or confuses that data with other information. Safety of flight can be compromised if you attempt to use the advanced avionics to substitute for required weather or aerodynamic needs. Safety of flight can be negated if you attempt to learn the advanced avionics system while in flight. You should use advanced avionics to reduce risk. Proper use of checklists and systematic training should be used to control common error-prone tasks and notice errors before they become a threat to safety of flight.

Chapter 2
Electronic Flight Instruments

Introduction

This chapter introduces the electronic flight instrument systems available with advanced avionics. You will see how electronic flight instrument systems integrate many individual instruments into a single presentation called a primary flight display (PFD). Since all flight instruments are combined in one integrated electronic flight instrument system, a number of enhancements to conventional flight instruments are now possible. In addition to learning to interpret the primary flight and navigation instruments, you must learn to recognize failures of the underlying instrument systems based on the indications you see in the cockpit. You must also maintain proficiency in using the backup/standby instruments that are still part of every advanced cockpit.

Primary Flight Display (PFD)

A PFD presents information about primary flight instruments, navigation instruments, and the status of the flight in one integrated display. Some systems include powerplant information and other systems information in the same display. A typical primary flight display is shown in *Figure 2-1*.

Primary Flight Instruments

Flight instrument presentations on a PFD differ from conventional instrumentation not only in format, but sometimes in location as well. For example, the attitude indicator on the PFD in *Figure 2-1* is larger than conventional round-dial presentations of an artificial horizon. Airspeed and altitude indications are presented on vertical **tape displays** that appear on the left and right sides of the primary flight display. The vertical speed indicator is depicted using conventional analog presentation. Turn coordination is shown using a segmented triangle near the top of the attitude indicator. The rate-of-turn indicator appears as a curved line display at the top of the heading/navigation instrument in the lower half of the PFD.

Cross-Checking the Primary Flight Instruments

The PFD is not intended to change the fundamental way in which you scan your instruments during attitude instrument flying. The PFD supports the same familiar control and performance, or primary and supporting methods you use with conventional flight instruments. For example, when using the primary and supporting method to maintain level flight, the altimeter is still the primary instrument for pitch, while the attitude indicator is a direct indicator and the vertical speed indicator provides supporting information. However, you need to train your eyes to find and interpret these instruments in their new formats and locations.

Common Errors: Altitude Excursions and Fixation

Pilots experienced in the use of conventional flight instruments tend to deviate from assigned altitudes during their initial experience with the PFD, while they adjust to the tape display presentation of altitude information. Another common error is the tendency to fixate and correct deviations as small as one to two feet at the expense of significant deviations on other parameters.

Figure 2-1. *A typical primary flight display (PFD).*

Enhancements to the Primary Flight Instruments

Some PFDs offer enhancements to the primary flight instruments. *Figure 2-2* shows an airspeed indicator that displays reference speeds (V-speeds) and operating ranges for the aircraft. Operating ranges are depicted using familiar color coding on the airspeed indicator. One negative human factor concerning this type of presentation should be remembered: while most of the displays are intuitive in that a high indication (such as climb pitch or vertical speed) is corrected by lowering the nose of the aircraft, the situation with the usual airspeed vertical tape is the opposite. In most current displays, the lower speeds are at the lower side of the airspeed indicator, while the upper or higher speeds are in the top portion of the airspeed display area. Therefore, if a low airspeed is indicated, you must lower the nose of the aircraft to increase, which is counterintuitive to the other indications.

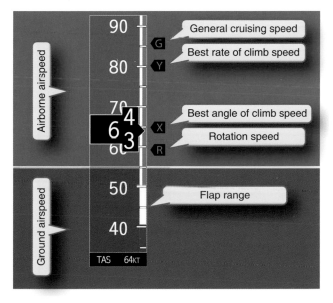

Figure 2-2. *Vertical airspeed (tape type) indicator.*

Figure 2-3 shows an attitude indicator that presents red symbols to assist in recovery from unusual attitudes. The symbols on the display recommend a lower pitch attitude.

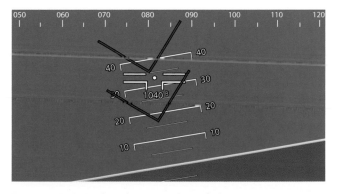

Figure 2-3. *Attitude indicator with symbols to assist in recovery from unusual attitude.*

Other valuable enhancements include trend indicators, which process data to predict and display future performance. For example, some systems generate "trend vectors" that predict the aircraft's airspeed, altitude, and bank angle up to several seconds into the future.

Primary Flight Instrument Systems

The primary flight instruments that appear on a PFD are driven by instrument sensor systems that are more sophisticated than conventional instrument systems. The attitude of the aircraft may be measured using microelectronic sensors that are more sensitive and reliable than traditional gyroscopic instruments. These sensors measure pitch, roll, and yaw movements away from a known reference attitude. Aircraft heading may be determined using a magnetic direction-sensing device such as a magnetometer or a magnetic flux valve.

Attitude and heading systems are typically bundled together as an attitude heading reference system (AHRS), which contains not only the sensors used to measure attitude and heading, but also a computer that accepts sensor inputs and performs calculations. Some AHRSs must be initialized on the ground prior to departure. The initialization procedure allows the system to establish a reference attitude used as a benchmark for all future attitude changes. As in any navigation system, attitude heading reference systems accumulate error over time. For this reason, AHRSs continually correct themselves, using periods of stable flight to make small corrections to the reference attitude. The system's ability to correct itself can be diminished during prolonged periods of turbulence. Some AHRSs can be reinitialized in flight, while others cannot. The pilot must become familiar with the operating procedures and capabilities of a particular system.

Information on altitude and airspeed is provided by sensors that measure static and ram air pressure. An air data computer (ADC) combines those air pressure and temperature sensors with a computer processor that is capable of calculating pressure altitude, indicated airspeed, vertical speed, and true airspeed. An air data attitude heading reference system (ADAHRS) combines all of the systems previously described into one integrated unit.

Navigation Instruments

PFDs and multi-function displays (MFDs) typically combine several navigation instruments into a single presentation. The instrument appearing at the bottom of the PFD in Figure 2-1 contains two navigation indicators: a course deviation indicator and a bearing pointer. These instruments can be displayed in a variety of views, and can be coupled to many of the navigation receivers (e.g., instrument landing system (ILS), global positioning system (GPS), very high frequency (VHF) omnidirectional range (VOR)) available in the aircraft. The pilot must, therefore, be sure to maintain an awareness of which navigation receivers are coupled to each navigation indicator.

MFDs may provide the same type of display as installed in the PFD position, but are usually programmed to display just the navigation information with traffic, systems data, radar Stormscope®/ Strikefinder®. However, in many systems, the MFD can be selected to repeat the information presented on the PFD, thereby becoming the standby PFD. The pilot should be absolutely certain of and proficient with the standby modes of operation.

More sophisticated PFDs present three-dimensional (3D) course indications. The primary flight display in *Figure 2-4* shows a 3D course indication, called a highway-in-the-sky (HITS) display. This display provides both lateral and vertical guidance along the planned flight path, while simultaneously presenting a 3D picture of the surrounding terrain. Keeping the symbolic aircraft within the green boxes on the display ensures that the flight remains within the selected GPS route and altitude. Consult the AFM and avionics manual for required navigational configuration for this function to be available.

Other Flight Status Information

An important feature of the PFD is its ability to gather information from other aircraft systems and present it to the pilot in the integrated display. For example, the PFD in *Figure 2-5* presents many useful items about the status of the flight. The top bar shows the next waypoint in the planned flight route, the distance and bearing to the waypoint, and the current ground track. The outside air temperature (OAT) is shown in the lower left corner of the display. The transponder code and status are shown with the current time in the lower right corner. This PFD also allows the pilot to tune and

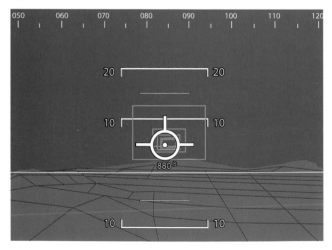

Figure 2-4. *An attitude indicator with HITS display symbology.*

identify communication and navigation radio frequencies at the top of the display.

Making Entries on the PFD

PFDs have evolved and have become more than flight displays in many cases. The amount of data available for display can overwhelm the pilot with data. Therefore, many manufacturers have integrated data control and display controls into the display unit itself, usually around the perimeter of the unit. These data and display controls provide different ways of selecting necessary information, such as altimeter settings, radials, and courses. *Figure 2-6* illustrates two different kinds of controls for making entries on primary flight displays. Some PFDs utilize a single knob and button-selectable windows to determine which entry is to be made. Other PFDs offer dedicated knobs for making entries; quantities are sometimes entered in one location and displayed in another. Still other units retain all controls on a separate control panel in the console or on the instrument panel.

Failures and the Primary Flight Display
Instrument System Failure

The competent pilot is familiar with the behavior of each instrument system when failures occur, and is able to recognize failure indications when they appear on the primary flight display. Manufacturers typically use a bold red "**X**" over, or in place of, the inoperative instruments and provide annunciator messages about failed systems. It is the pilot's job to interpret how this information impacts the flight.

The inoperative airspeed, altitude, and vertical speed indicators on the PFD in *Figure 2-7* indicate the failure of the air data computer. As do all electronic flight displays, navigation units (area navigation (RNAV)/flight management systems (FMS)) and instrumentation sensors rely on steady, uninterrupted power sources of 24 VDC or 12 VDC power. Any interruptions in the power supplies, such as alternator/ regulator failure, drive belt failure, lightning strikes, wiring harness problems, or other electrical failures, can completely disrupt the systems, leading to erratic indications or completely inoperative units. Especially in standard category aircraft not designed or built with the redundancy inherent in transport category aircraft, a proficient and prudent pilot plans for failures and has alternate plans and procedures readily available.

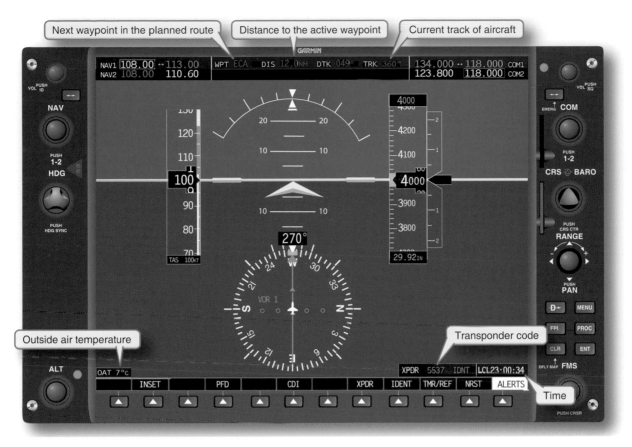

Figure 2-5. *PFD flight status items.*

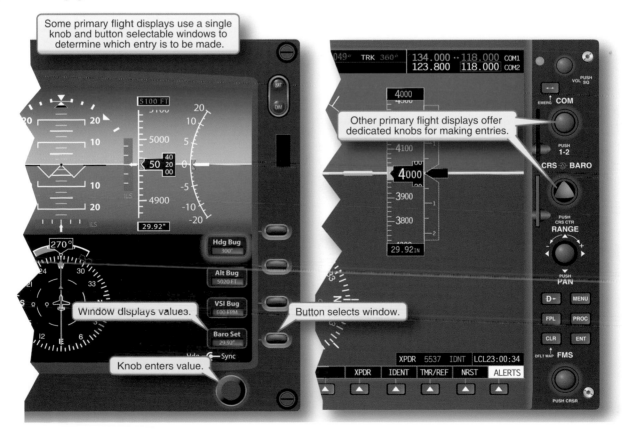

Figure 2-6. *Making entries on a PFD.*

Figure 2-7. *A PFD indicating a failed air data computer.*

The inoperative attitude indicator on the PFD in *Figure 2-8* indicates the failure of the AHRS. By understanding which flight instruments are supported by which underlying systems (e.g., ADC, attitude heading reference system (AHRS)), you can quickly understand the source of a failure. It is important to be thoroughly familiar with the operation of the systems and the abnormal/emergency procedures in the pilot's operating handbook (POH), aircraft flight manual (AFM), or avionics guides.

Figure 2-8. *A PFD indicating a failed AHRS.*

PFD Failure

The PFD itself can also fail. As a first line of defense, some systems offer the reversion capability to display the PFD data on the multi-function display (MFD) in the event of a PFD failure.

Every aircraft equipped with electronic flight instruments must also contain a minimal set of backup/standby instruments. Usually conventional "round dial instruments," they typically include an attitude indicator, an airspeed indicator, and an altimeter. Pilots with previous experience in conventional cockpits must maintain proficiency with these instruments; those who have experience only in advanced cockpits must be sure to acquire and maintain proficiency with conventional instruments.

Awareness: Using Standby Instruments

Because any aircraft system can fail, your regular proficiency flying should include practice in using the backup/standby instrumentation in your aircraft. The backup/standby instrument packages in technically advanced aircraft provide considerably more information than the "needle, ball, and airspeed" indications for partial panel work in aircraft with conventional instrumentation. Even so, the loss of primary instrumentation creates a distraction that can increase the risk of the flight. As in the case of a vacuum failure, the wise pilot treats the loss of PFD data as a reason to land as soon as practicable.

Essential Skills

1. Correctly interpret flight and navigation instrument information displayed on the PFD.

2. Determine what "fail down" modes are installed and available. Recognize and compensate appropriately for failures of the PFD and supporting instrument systems.

3. Accurately determine system options installed and actions necessary for functions, data entry and retrieval.

4. Know how to select essential presentation modes, flight modes, communication and navigation modes, and methods mode selection, as well as cancellation.

5. Be able to determine extent of failures and reliable information remaining available, to include procedures for restoring function(s) or moving displays to the MFD or other display.

Chapter Summary

The primary flight instruments can all be displayed simultaneously on one reasonably easy-to-read video monitor much like the flat panel displays in laptop computers. These displays are called primary flight displays (PFDs). You must still cross-check around the panel and on the display, but more information is available in a smaller space in easier to read colors. These convenient displays receive data from sensors such as magnetometers or magnetic flux valves to determine heading referenced to magnetic north. The attitude (pitch and roll) of the aircraft is sensed by the attitude heading reference system (AHRS) and displayed as the attitude gyro would be in conventional instrumentation. The altitude, airspeed, and outside temperature values are sensed in the air data computer (ADC) and presented in the PFD on vertical scales or portions of circles.

The multi-function display (MFD) can often display the same information as the PFD and can be used as a backup PFD. Usually the MFD is used for traffic, route selection, and weather and terrain avoidance. However, some PFDs also accommodate these same displays, but in a smaller view due to the primary flight instrument areas already used in the display. You must learn and practice using that specific system.

It is important to be very careful in the selection (programming) of the various functions and features. In the event of failures, which have a large impact on flight safety and situational awareness, you must always be ready and able to complete the flight safely using only the standby instruments.

Chapter 3
Navigation

Introduction

This chapter introduces the topic of navigation in the advanced cockpit. You will learn about flight management systems (FMS) and area navigation (RNAV) systems, an increasingly popular method of navigating that allows pilots to make more efficient use of the national airspace system. The increasing number of users is attributable to more economical and accurate satellite signal receivers and computer chips. RNAV systems may use VHF omnidirectional range (VOR); distance measuring equipment (DME) (VOR/DME, DME/DME) signals; inertial navigation systems (INS); Doppler radar; the current version of LOng RAnge Navigation (LORAN), LORAN-C (and eLORAN, as it becomes operational); and the global positioning system (GPS), to name a few. Ground-based LORAN-C is a reliable complement to space-based GPS systems (United States Department of Defense (DOD) GPS, Russian Global Navigation Satellite System (GLONASS), and European Galileo in the future).

Wide area augmentation system (WAAS) of the standard GPS furnishes additional error correction information, allowing Category I precision approaches (similar to basic instrument landing system (ILS) minimums) to units equipped to receive and integrate the data. Most general aviation pilots learn to work with an FMS unit primarily using GPS signals, possibly with WAAS and LORAN-C options. Older RNAV units made use of VOR and DME information to compute positions within range of the navaids. Newer units contain databases that allow route programming with automatic sequencing through the selected navigation points. Therefore, flight management system (FMS) is the best descriptor of the current GPS units integrating VOR (and DME, optionally) to allow point-to-point navigation outside established flight routes. You will learn to use the FMS data entry controls to program a flight route, review the planned route, track and make modifications to the planned route while en route, plan and execute a descent, and fly an approach procedure that is based solely on RNAV signals. You should remember that FMS/RNAV units requiring external signals for navigation are usually restricted to line-of-sight reception (LORAN-C being somewhat of an exception). Therefore, navigation information in valleys and canyons that could block satellite signals may be severely restricted. Users in those areas should pay particular attention to the altitude or elevations of the satellites when depending on space-based signals and plan flight altitudes to ensure line-of-sight signal reception. Review the GPS unit's documentation sufficiently to determine if WAAS is installed and how WAAS corrections are indicated.

You will learn how the FMS can automatically perform many of the flight planning calculations that were traditionally performed by hand, and the importance of keeping flight planning skills fresh. You will also discover how the FMS can help you detect and correct errors made in the flight planning process, how the complexities of the FMS make some new kinds of errors possible, and techniques to help avoid them.

Last, you will see how advanced cockpit systems can be used to navigate using ground-based navigation facilities such as VOR and DME. Maintaining pilot skills using ground-based navigation facilities is a simple matter of occasionally using them as the primary means of navigation, and as a backup to verify position and progress when RNAV is used.

Area Navigation (RNAV) Basics

RNAV Concept

Area Navigation (RNAV) is a navigation technique that allows pilots to navigate directly between any two points on the globe. Using RNAV, any location on the map can be defined in terms of latitude and longitude and characterized as a waypoint. Onboard RNAV equipment can determine

the present position of the aircraft. Using this positional information, the equipment can calculate the bearing and distance to or from any waypoint and permit navigation directly between any two waypoints. In this way, RNAV overcomes a fundamental limitation of conventional navaid point-to-point navigation techniques, which require navigating between electronic navigation transmitters on the ground. The following examples illustrate this limitation.

An aircraft equipped with conventional VOR receivers is positioned at Point A as shown in the diagram at the top of *Figure 3-1,* and the pilot wishes to navigate directly to Point B. Although there appear to be a few VOR stations in the vicinity of the aircraft, it is not clear whether reception is possible from the aircraft's present position. If the VOR stations are within reception range, the pilot has two choices: (1) fly to intercept the closest airway, then track it to the intersection; or (2) fly to intercept an extension of the radial that defines Point B (assuming reception is possible). Neither alternative provides the pilot with a means of flying directly to the intersection.

Suppose the same aircraft is positioned at Point A as shown at the bottom of *Figure 3-1* and the pilot wishes to navigate directly to Point C, which is neither a VOR station nor airway intersection. This pilot has an even more difficult situation. Assuming the VOR stations are within reception range, the pilot needs to create two makeshift airways using a navigation plotter and chart, fly to intercept one of them, then track to Point C (which the pilot has defined as the intersection between the two courses). Flying a direct course to Point C with any degree of accuracy is not possible. Since RNAV systems are not bound by these limitations, the entire airspace is available for navigational use. The national airspace system can thus accommodate more aircraft. However, when the pilot leaves the established airways, he or she also leaves the guaranteed obstruction clearances designed into the airway system. Always plan flights above the maximum elevation figure (MEF) displayed on sectional charts when flying off airways, and be aware that manmade obstructions such as towers may not be added to charts for some time after construction. If flying a new routing, allow for construction, which may not be published yet.

FMS/RNAV Computer

RNAV is possible through use of a variety of navigation facilities and installed aircraft equipment operated in the U.S. National Airspace System. This handbook focuses on the more common GPS RNAV, a satellite-based radio navigation system available to aircraft equipped with a GPS receiver. In addition to its ability to receive signals from GPS satellites, a GPS receiver also contains a computer processor and a navigation database that includes much of the

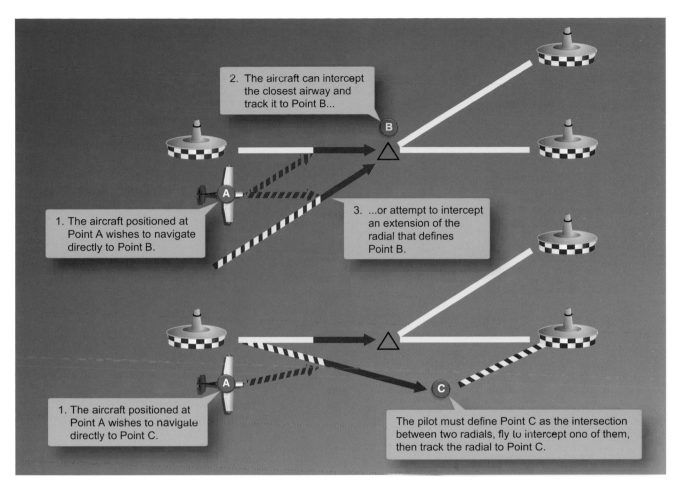

Figure 3-1. *Limitations of conventional navigation.*

information found on en route and terminal procedure charts. The newer, more capable units provide map displays, traffic and weather overlays of data, contain VOR/DME/localizer/glideslope receivers, and can compute fuel usage in addition to the navigation route information. For this reason, the more descriptive term "FMS" is used in this handbook to refer to these GPS receivers.

An FMS allows you to enter a series of waypoints and instrument procedures that define a flight route. If these waypoints and procedures are included in the navigation database, the computer calculates the distances and courses between all waypoints in the route. During flight, the FMS provides precise guidance between each pair of waypoints in the route, along with real-time information about aircraft course, groundspeed, distance, estimated time between waypoints, fuel consumed, and fuel/flight time remaining (when equipped with fuel sensor(s)).

FMS/RNAV/Autopilot Interface: Display and Controls

Every avionics device has a display and a collection of buttons, keys, and knobs used to operate the unit. The display allows the device(s) to present information. The controls

allow the pilot to enter information and program the avionics to accomplish the desired operations or tasks. The display and controls for a typical FMS are shown in *Figure 3-2*.

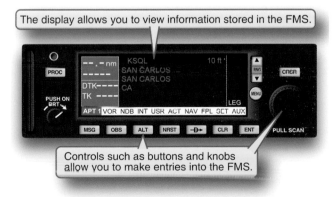

Figure 3-2. *FMS display and controls.*

Accessing Information in the FMS

FMS units contain much more information than they can present on the display at any one time. Information pertaining to some topics often extends beyond what can be presented on a single page. Page groups, or chapters, solve this problem by collecting all of the pages pertaining to the same topic. Each

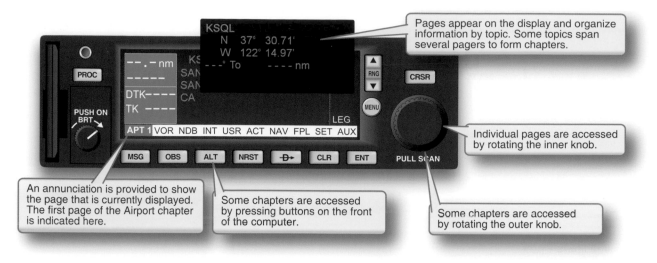

Pages appear on the display and organize information by topic. Some topics span several pagers to form chapters.

Individual pages are accessed by rotating the inner knob.

An annunciation is provided to show the page that is currently displayed. The first page of the Airport chapter is indicated here.

Some chapters are accessed by pressing buttons on the front of the computer.

Some chapters are accessed by rotating the outer knob.

Figure 3-3. *Pages and page groups (chapters).*

page presents information about a particular topic, and bears a page title reflecting its content. For example, the airport chapter may be divided into several airport pages, each page displaying different information about that airport. One page might be navaids. Another page might be the airport taxiway diagram. Yet another airport page might indicate available services and fixed-base operators. Review the documentation for that specific unit and installation to determine what information and levels of data are available and require updates. Usually, only one page can be displayed at a time. The airport page is displayed on the FMS in *Figure 3-3*.

Figure 3-3 shows how to access pages and chapters on one manufacturer's FMS. Different FMS units have different ways of allowing the pilot to switch between chapters and pages, and different ways of informing the pilot which chapter and page is currently displayed.

Making Entries in the FMS

To enter data, you use the FMS buttons (keyboard or individual) and knob controls, or a data source, such as disk media or keypad, as shown in *Figure 3-4*.

FMS units that do not feature keypads typically require the pilot to make entries using the same knobs to move among chapters and pages. In this case, the knobs have multiple purposes and, thus, have different modes of operation. To use the knobs for data entry, you must first activate what some manufacturers call the "cursor" (or "data entry") mode. Activating the cursor mode allows you to enter data by turning the knob. In other units, after activating the data entry mode, entries are made by pushing buttons.

Figure 3-5 illustrates the use of cursor mode to enter the name of an airport using one FMS. Pressing the inner knob engages cursor mode. A flashing cursor appears over one of

Figure 3-4. *An FMS keypad.*

the items on the page, indicating that it is ready for editing. Then, the inner knob is used to dial letters and numbers; the outer knob is used to move the flashing cursor between items on the page.

Integrated Avionics Systems

Some systems integrate FMS/RNAV display and controls into existing cockpit displays usually called PFDs and MFDs. In this case, there is no separate display to point to and call the RNAV display. *Figure 3-6* shows a system that uses the PFD to provide controls and a display for the FMS. This type of system utilizes the same concepts and procedures described above to access and enter into the navigation computer.

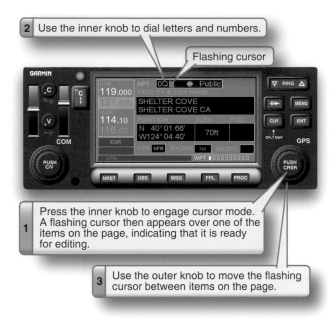

2 Use the inner knob to dial letters and numbers.

Flashing cursor

1 Press the inner knob to engage cursor mode. A flashing cursor then appears over one of the items on the page, indicating that it is ready for editing.

3 Use the outer knob to move the flashing cursor between items on the page.

Figure 3-5. *Making entries using cursor mode.*

Learning: Simulators for Learning and Practice

Avionics simulators can assist the pilot in developing proficiency in the advanced cockpit. Some manufacturers offer computer-based simulators that run on a personal computer and let the pilot learn how the unit organizes and presents information, as well as practice the button-pushing and knob-twisting procedures needed to access and enter data. One very important function that every pilot of programmable avionics should learn and remember is how to cancel entries and functions. Turbulent flight conditions make data entry errors very easy to make. Every pilot should know how to revert quickly to the basic aircraft controls and functions to effect recovery in times of extreme stress. These programs are extremely useful not only for initial learning, but also for maintaining proficiency. For more sophisticated training, many manufacturers of flight simulators and flight training devices are now developing devices with advanced cockpit systems. These training platforms allow the pilot to work through realistic flying scenarios that teach not only the operating procedures required for each system, but also how to use the systems most effectively.

Flight Planning

Preflight Preparation

Title 14 of the Code of Federal Regulations (14 CFR) part 91, section 91.103 requires you to become familiar with all available information before beginning a flight. In addition to the required checks of weather, fuel, alternate airports, runway lengths, and aircraft performance, there are a number of requirements unique to the use of avionics equipment. Many of these considerations apply specifically to the use of FMS/RNAV under instrument flight rules (IFR). However,

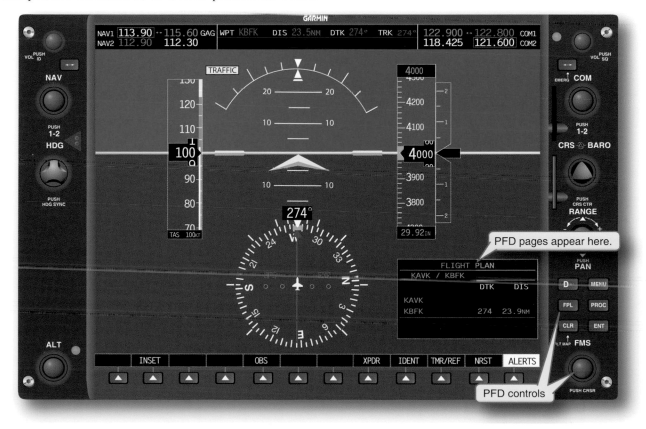

PFD pages appear here.

PFD controls

Figure 3-6. *An integrated avionics system.*

a check of these same requirements before operating under visual flight rules (VFR) enhances safety and enforces good habit patterns, which have been proven to greatly enhance aviation safety.

FMS/RNAV Approval for IFR Operations

Only some FMS/RNAV units are approved for IFR navigation, and it is important to make this determination before flying with any particular unit. Sometimes, this limitation is based on the installation (i.e., method of installation, qualifications of installer), aircraft approval, availability of approved maintenance, and geographic location. No hand-held GPS unit is approved for IFR navigation, and many panel-mounted units are restricted to VFR use only.

Even when an FMS is approved for IFR, the installation of the system in that specific aircraft must also be approved. Even if you have an IFR-approved FMS unit, you may not use it for IFR navigation unless the installation is approved as well. This approval process usually requires a test flight to ensure that there are no interfering inputs, signals or static emanating from the aircraft in flight. RNAV units that do not meet all of these requirements may still be used as situation enhancing navigation resources when operating under instrument flight rules.

The first place to check when determining IFR certification for an FMS is the Pilot's Operating Handbook (POH) or Aircraft Flight Manual (AFM). For every aircraft with an IFR approved FMS/RNAV unit, the AFM explicitly states that the unit has been approved for IFR navigation and what IFR operations are specifically authorized for that installation.

Navigation Database Currency

The navigation database contained in the FMS must be current if the system is to be used for IFR navigation and approaches. Some units allow en route IFR operations if the navigation waypoints are manually verified by the pilot and accepted. The effective dates for the navigation database are shown on a start-up screen that is displayed as the FMS cycles through its startup self-test. Check these dates to ensure that the navigation database is current. *Figure 3-7* shows the start-up screen and effective dates for one popular FMS.

Alternative Means of Navigation

To use some GPS-based RNAV units (those certified under Technical Standard Order (TSO) 129) for IFR flight, an aircraft must also be equipped with an approved and alternate means of IFR navigation (e.g., VOR receiver) appropriate to the flight. Ensure that this equipment is onboard and operational, and that all required checks have been performed (e.g., 30-day VOR check).

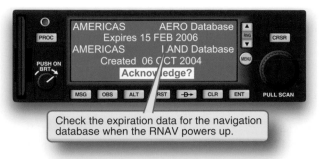

Figure 3-7. *Checking the navigation database.*

The avionics operations manual/handbook should state the certification status of the installed system. The supplements to the AFM should state the status of the installed equipment, including the installed avionics. Most systems require that the advanced avionics manuals be on board as a limitation of use.

NOTAMs Relevant to GPS

There are numerous notices to airmen (NOTAMs) that apply specifically to users of navigation aids. For example, when anomalies are observed in the behavior of the global positioning system, or when tests are performed, a GPS UNRELIABLE NOTAM is issued. Similarly, published instrument procedures that rely on RNAV equipment sometimes become "Not Available" when safety concerns arise, such as ground-based interference. It is important to check all NOTAMs prior to IFR flights and, especially, GPS and WAAS NOTAMs before flying. Remember, when talking to a flight service station (FSS)/automated flight service station (AFSS) briefer, you must specifically request GPS/WAAS NOTAMs.

GPS Signal Availability

GPS-based RNAV equipment that uses the DOD GPS relies on adequate signal reception throughout the course of a flight. Signal reception becomes especially critical during instrument approaches when signal reception criteria become more stringent. Signal reception is generally predictable, and you can request information on likely signal reception for the destination airport in the preflight briefing from Flight Service. Many GPS RNAV units include a feature called receiver autonomous integrity monitoring (RAIM) that allows you to view predictions about future signal reception at specific locations. WAAS-enabled receivers do not have this restriction or limitation due to the error corrections available from the WAAS. WAAS is a form of differential GPS (DGPS) providing enhanced position accuracy. Each Wide Area Reference Station (WRS) provides correction data to a Wide Area Master Station (WMS), which computes a grid of correction data to be uplinked to a geostationary satellite (GEO) from a Ground Earth Station (GES). The geostationary satellite transmits the correction data (and also navigation data) to the user on the L1 GPS navigation frequency (1575.42 MHz). The user GPS receiver uses the downlink WAAS data to correct received navigation data. The goal of WAAS is to obtain at least a 7-meter horizontal and vertical accuracy.

Local Area Augmentation System (LAAS), when it becomes available, is another DGPS mode which is designed to provide 1-meter accuracy for precision approaches. It uses a local error VHF transmitter near the runway providing a direct link from the sensor to the aircraft GPS receiver.

Alternate Airports

It is very important to know what equipment is installed in the aircraft. GPS-based FMS/RNAV units certified to TSO-C145A or TSO-146A may be used when an alternate airport is required in the flight plan for the approaches at the destination and alternate airport if the WAAS is operational. No other navigation avionics would be required. Units certified under TSO-C129 are not authorized for alternate approach requirements. The aircraft must have stand-alone navigation equipment, such as VOR, and there must be an approved instrument approach at the alternate airport based on that equipment. (However, once diverted to the alternate airport, the pilot could fly a GPS-based approach there, as long as there is an operational, ground-based navaid and airborne receiver in the aircraft for use as a backup.)

Aircraft Equipment Suffixes

Since air traffic control (ATC) issues clearances based on aircraft equipment suffixes, consult the Aeronautical Information Manual (AIM) Table 5-1-2, Aircraft Suffixes, to ensure that the flight plan includes the correct equipment suffix for a particular aircraft. Use the suffix that corresponds to the services and/or routing that is needed. For example, if the desired route or procedure requires GPS, file the suffix as "/G" or "/L," as appropriate to that aircraft, and operational equipment installed. (Remember that minimum equipment list (MEL) deferred items can change the status of the aircraft.)

Suitability of an RNAV Unit for VFR Flight

Even when an RNAV receiver is to be used only for supplemental ("supplemental" meaning a situation enhancing source of navigation information, but not the primary or sole source of navigation information) navigation information during VFR flight, you should consider these suitability factors in the interest of safety. The use of an expired navigation database might cause you to stray into airspace that was not yet designated at the time the expired navigation database was published. Some VFR-only GPS units do not alert you when signal reception has faded, which could lead to reliance on erroneous position information. Lack of attention to the "see and avoid" basic principle of every visual meteorological conditions (VMC) flight means too much time spent focused inside the cockpit on advanced avionics versus staying synchronized with the flight events, possibly creating a life-threatening total flight situation.

Programming the Flight Route

The procedures used to program an FMS with your intended route of flight are fundamentally the same in all types of systems, yet many differences are evident. The primary difference between systems lies mainly in the "knob or switchology"—the specific design features, operational requirements, and layout of the controls and displays used to operate the avionics. Be thoroughly familiar with the procedures required for each FMS or RNAV unit to be used.

Suppose you have planned a flight from San Carlos Airport (KSQL) to Oakdale Airport (O27), as shown in the flight plan appearing in *Figure 3-8*. The planned route proceeds directly to SUNOL intersection, then follows V195 until reaching ECA, the initial approach fix for the GPS Runway 10 approach into Oakdale. The distances, bearings, estimated times en route, and fuel requirements for the flight have all been calculated. The next step is to enter some of these details into the FMS.

NAVIGATION LOG								
Aircraft Number: **N1361M**	Dep: **KSQL**		Dest: **027**			Dest: **11/06/06**		
Clearance: **C – 027**								
R – Direct SUNOL, V195 ECA, Direct 027								
A – CLB 5000								
F – 12L3								
T – 0356								
Estimated Time En Route = 0.49								
Check Points (Fixes)	Ident.	Course Route	Altitude	Mag Crs.	Fuel	Dist.	GS	Time Off
	Freq.				Leg Rem.	Leg Rem.	Est. Act.	ETE ETA ATE ATA
KSQL TWR 119.0	DEP 121.3				48	78.6		ETE ETA ATE ATA
SUNCL	SJC 1141	060	5000		21 45.9	21.5 57.1	81	016
TRACY	ECA 116.0	049	5000		1.2 44.7	18 39.1	120	0.09
ECA (IAF)	ECA 116.0	049	3000		1.1 43.6	15 24.1	120	0.08
MOTER	ECA 116.0	084	2000		0.7 42.9	8 16.1	1000	0.05
ZOSON (FAF)	ECA 116.0	084	2000		0.6 42.3	55 10.6	90	0.04
RW 10 (MAP)	ECA 116.0	096	MDA 720		0.4 41.9	4.5 61	90	0.03
WRAPS (HOLD)	LIN 114.8	352	3000		0.6 41.3	61 0	90	0.04

Figure 3-8. *A conventional flight plan.*

The Flight Planning Page

Every FMS unit includes a page dedicated to entering a flight plan. Typically, entering a flight plan is a simple matter of "filling in the blanks"—entering the en route waypoints and instrument procedures that make up the planned route.

En Route Waypoints and Procedural Waypoints

Entering a flight route into the FMS unit requires you to enter the waypoints that define your route. FMS distinguish between two kinds of waypoints: (1) waypoints that are published, such as departure, arrival, or approach procedure points; and (2) user defined waypoints. The approved system software (the internal programming) allows the pilot to manually enter airport and en route waypoints. However, you are prohibited by the software from entering (or deleting) individual waypoints that define a published instrument procedure, since misspelling a procedural waypoint name or deleting a procedural waypoint (e.g., final approach fix) could have disastrous consequences. Any changes to the selected database approach procedure will cancel the approach mode. Changing to go direct to a waypoint will not, in most units, cancel the approach mode (such as receiving radar vectors to final and bypassing an intermediate fix).

Entering En Route Waypoints

Looking at the planned route in *Figure 3-8,* it is apparent that San Carlos airport (KSQL), and SUNOL and TRACY intersections are not part of any instrument procedure that pertains to the planned flight. These waypoints can be entered into the unit, as shown in *Figure 3-9.*

The remaining waypoints in *Figure 3-8,* starting with the initial approach fix at ECA, are part of the Oakdale GPS approach procedure. Waypoints that are part of a published instrument procedure are entered by a different technique that will be introduced later. In some cases, you navigate along an airway that contains a string of waypoints, such as the one shown in *Figure 3-10.*

In this case, it is only necessary to enter waypoints along the airway that represent course changes. In *Figure 3-10,* REANS intersection is a changeover point that joins the

Figure 3-9. *Entering en route waypoints in the flight plan.*

Pomona 073-degree radial and the Twentynine Palms 254-degree radial. For this airway segment, you could enter POM, REANS, and TNP, keeping in mind that the remaining waypoints do not appear in the programmed route.

Entering Airways

More sophisticated FMSs allow you to enter entire airways with a single action into the unit. When an airway and

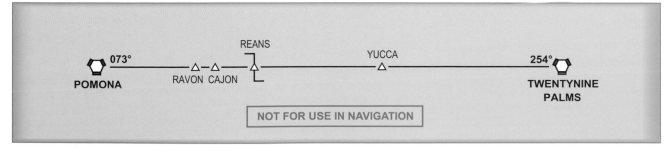

Figure 3-10. *Entering waypoints along an airway.*

endpoint for that airway are selected, all waypoints that occur along the airway are automatically inserted into the flight plan. *Figure 3-11* shows a navigation unit that allows airways to be selected.

Entering Procedures

Every IFR-capable FMS offers a menu of published instrument procedures, such as departures, arrivals, and approaches. When you choose one of these procedures, the FMS automatically inserts all waypoints included in that procedure into the flight plan. *Figure 3-12* illustrates how you might choose an approach procedure using one popular FMS.

Risk: Taking Off Without Entering a Flight Plan

The convenience of the FMS, especially the "direct to" feature common to all units, creates the temptation to program only the first en route waypoint prior to takeoff and then enter additional waypoints once airborne. Keep in mind, however, that no matter how skilled you become with the avionics, programming requires "heads down" time, which reduces your ability to scan for traffic, monitor engine instruments, etc. A better strategy is to enter all of the flight data before you take off.

Reviewing the Flight Route

Once a route has been entered into the FMS, the next step is to review the route to ensure it is the desired route. It is particularly important to ensure that the programmed route agrees with the pilot's clearance, the en route and terminal area charts, and any bearing, distance, time, and fuel calculations that have been performed on paper.

Catching Errors: Using the FMS Flight Planning Function To Cross-Check Calculations

Using the FMS's flight planning function to check your own flight planning calculations is one example of how advanced cockpit systems can decrease the likelihood of an error. Alternatively, cross-check the computer's calculations against your own. (Remember the old computer programmer's adage, "Garbage in, garbage out (GIGO).")

The flight planning page can also be used to review the route that you entered into the FMS, as illustrated in *Figure 3-13*. Be sure to check at least four things when reviewing your route.

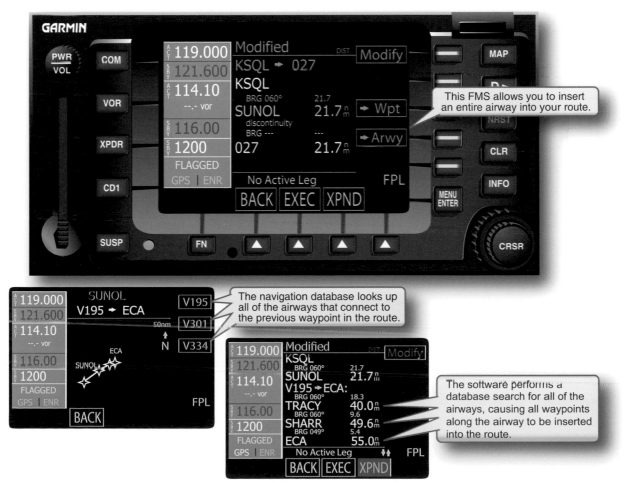

Figure 3-11. *Inserting an airway into a flight route.*

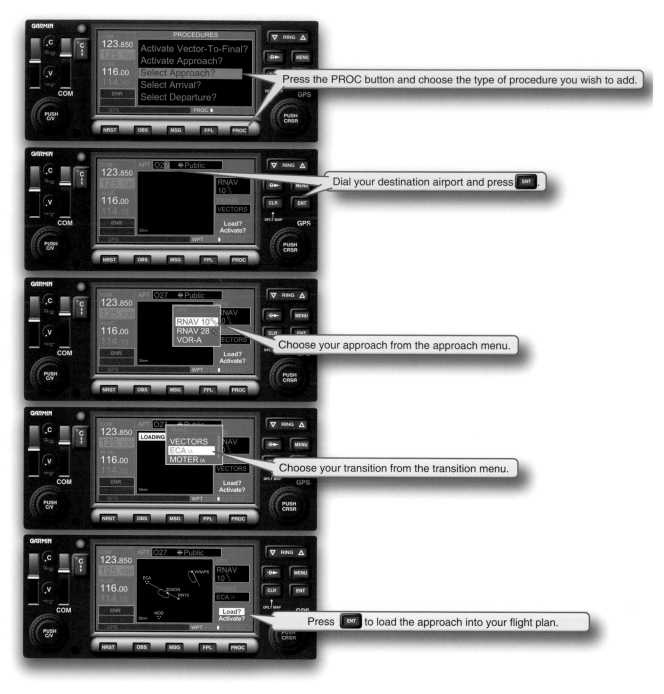

Figure 3-12. *Inserting published instrument procedures into a flight route.*

Check the Waypoints

On the flight planning page, compare the sequence of waypoints with that prescribed by his or her clearance. Are any waypoints missing? Did you mistakenly include any extra waypoints in the route? Did you misspell any waypoints? Did the computer mistakenly include any extra waypoints in the route?

Check the Distances

On the flight planning page, you can see that the computer has calculated the distances between the waypoints in the route. These distances can be checked against the en route charts. A common error is to misspell the name of a waypoint and, thus, mistakenly enter a waypoint not appropriate to the planned route (e.g., KHEE versus KHEF). Checking the waypoint distances for unusual numbers is a good way to spot these errors.

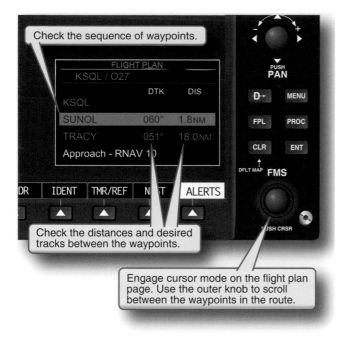

Check the sequence of waypoints.

Check the distances and desired tracks between the waypoints.

Engage cursor mode on the flight plan page. Use the outer knob to scroll between the waypoints in the route.

Figure 3-13. *Reviewing the flight route.*

Check the Desired Tracks

On the flight planning page, you can also see the course that the computer has calculated between waypoints along the route. A desired track between two waypoints represents the shortest path between them. The desired track between two waypoints may differ from the course seen on the aeronautical charts. In fact, there may be a difference of several degrees between the desired track and the airway course. Some of this difference may be due to the method in which the FMS accounts for magnetic variation. Some units use an internal database and interpolate, while others compute all values from tables.

Unlike the world as printed on paper charts, the earth is round, not flat. The shortest distance between two points on the earth is not a straight line; it is an arc, as shown in *Figure 3-14.*

The shortest route between two points on the surface of the earth is called a great circle route. To find the great circle

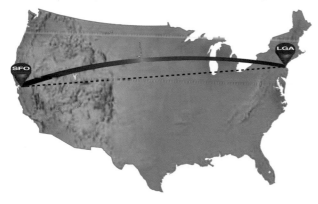

Figure 3-14. *A great circle route.*

route that connects two points, imagine a geometric plane cutting through the earth that passes through the two points and the center of the earth.

On the great circle route from SFO to LGA in *Figure 3-14,* departing SFO, the desired track is a little less than 90 degrees. Upon arrival at LGA, it appears to be greater than 90 degrees. The desired track heading is constantly changing since it is a circle, not a line. If, however, the difference exceeds several degrees, you need to investigate further to determine the cause.

Check for Route Discontinuities

Some FMS units do not automatically assume that you wish to fly between each of the waypoints that have been entered into the flight plan. When there is a question about how to proceed from one waypoint or instrument procedure to the next, some units insert a "discontinuity" in the programmed route. A route discontinuity indicates that the FMS needs further input from you about how two route segments should be connected. A route discontinuity is shown in *Figure 3-15.* If you wish to proceed directly from the waypoint that appears before the route discontinuity to the waypoint that appears after, you can simply delete the discontinuity, as shown in *Figure 3-15.*

If the route discontinuity is left in the flight plan, the unit computer will not provide guidance beyond the waypoint that occurs before the discontinuity.

Maintaining Proficiency: Aeronautical Knowledge

It is easy to use an FMS without performing your own calculations for courses, headings, times, distances, and fuel used, but studies have demonstrated that aeronautical skills that are not practiced regularly quickly fade, regardless of experience level or certificates and ratings held. Abnormal and emergency situations (e.g., electrical failure) do occur, so it is important to maintain proficiency in at least making "rule of thumb" calculations on your own.

Coupling the FMS to the Navigation Indicator(s)

Every advanced avionics cockpit features one or more navigation instruments used for course guidance. The navigation indicator (e.g., a horizontal situation indicator (HSI) or electronic HSI) may include one or more course deviation indicators (CDIs), as well as one or more radio magnetic indicators (RMIs). When automatic course/ en route/ approach tracking is desired, you must couple (or connect) the FMS to the autopilot and select "navigation" as the source for the autopilot versus "heading" source, for example. With VOR navigation, that was sufficient. Now, with multiple sources of navigation data available, you must also ensure that the proper navigation information source

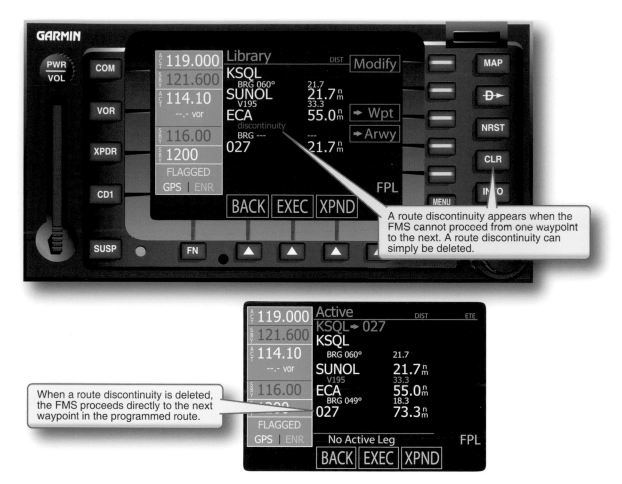

Figure 3-15. *A route discontinuity and deletion.*

was selected in the FMS. Every advanced cockpit offers buttons or switches that allow you to choose which navigation indications will be shown on which display or instrument.

This situation becomes complicated in aircraft that contain dual FMS/RNAV installations and redundant selectable displays or instruments. The pilot must learn how to configure each navigation instrument to show indications from each possible navigation source.

Figure 3-16 shows an example of a primary flight display (PFD) navigation indicator that combines a course deviation indicator (CDI) and a radio magnetic indicator (RMI), and allows the pilot to display indications from one of two FMS on either indicator.

Common Error: Displaying the Wrong Navigation Source

The annunciations showing which navigation sources are displayed on which navigation instruments are often small, so there is significant potential for displaying a navigation source other than the one you intended to select. The consequences of losing track of which navigation signals you are following

can be significant: you may think you are steering along one course when in fact you are steering along a different one. Be sure to verify these settings prior to departure, and again each time you make changes to any navigation instrument. Some installations compound this potential with automatic source switching. The most common switching mode is a GPS source to be automatically deselected when the VOR is set to an ILS localizer frequency and a signal is present. Typically, that is not a problem since the pilot intends to switch to the ILS anyway. However, the error arises upon missed approach, when the pilot selects another frequency to follow a VOR missed approach routing. At that point, some units revert back to the previous GPS or other RNAV routing selected instead of the VOR frequency that the pilot just picked. This can result in gross navigation errors and loss of obstruction clearances. In some units, this is a shop programmable or jumper selected option. Check your unit's features. Always check for correct navigation source selection and cross-check against the published procedure. Be ready and able to fly and navigate manually.

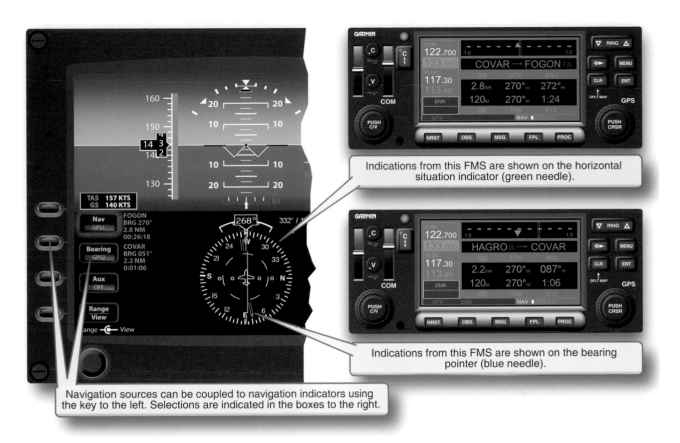

Figure 3-16. *Coupling the FMS to navigation instruments.*

Awareness: Mode Awareness

Mode awareness refers to the pilot's ability to keep track of how an advanced avionics cockpit system is configured. As shown in *Figure 3-16*, every advanced avionics system offers an annunciation of which mode is currently set—somewhere in the cockpit! There is no guarantee that you will notice these annunciations in a timely manner. The configuration of these systems must remain part of your mode situational awareness at all times. One strategy is to include "mode checks" as part of your checklist or callout procedures. For example, after programming a route into the FMS, verify that the navigation indicator shows course guidance from the desired source, and that the indication agrees with your estimate of the correct direction and distance of flight.

Essential Skills

1. Determine whether the FMS is approved for the planned flight operation.

2. Determine if your FMS can be used as a primary navigation system for alternate requirements.

3. Understand how entries are made and how the entries can be canceled.

4. Understand how that unit(s) is installed, and how it is programmed or jumpered for optional functions.

5. Determine which navigation sources are installed and functional.

6. Determine the status of the databases.

7. Program the FMS/RNAV with a flight plan, including en route waypoints, user waypoints, and published instrument procedures.

8. Review the programmed flight route to ensure it is free from error.

9. Find the necessary pages for flight information in the databases.

10. Determine which sources drive which displays or instruments, and where the selection controls are located.

11. Determine and understand how to use and program optional functions and equipment installed with FMS/RNAV basic unit.

En Route Navigation

The FMS provides guidance toward each waypoint in the programmed flight route, and provides information to help you track your progress.

The Active Waypoint

In normal navigation, at any given time, the aircraft is progressing to the next waypoint in the programmed flight route. This next waypoint is called the active waypoint. FMSs typically display the active waypoint on a page dedicated to showing flight progress. While "going to" is the normal function for navigation, nearly all FMSs have the provision to select a point, waypoint, or navaid to navigate "from" that point or position. This can be useful for holding, tracking NDB bearing, grids, etc., and allows the tracking of a bearing with the autopilot engaged and coupled to a navigation source. However, if you are doing an ADF approach, the primary navigation source must be available to support that approach. The page shown in *Figure 3-17* indicates that TRACY is the active waypoint. The primary flight display in *Figure 2-5* shows the active waypoint (ECA) at the top of the display.

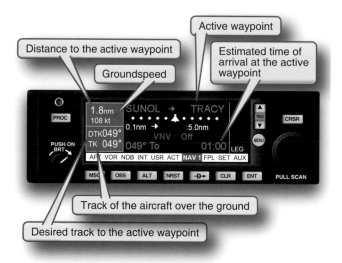

Figure 3-17. *Active waypoint, desired track, track, and ETA at active waypoint.*

Desired Track

The FMS/PFD/MFD navigation display also shows the desired track to the active waypoint. The desired track is the intended course for the active leg in the programmed flight plan. It is the track that connects the waypoint the aircraft just passed to the active waypoint. On the display in *Figure 3-17,* the current desired track is the 049-degree course between the SUNOL and TRACY waypoints.

Track

The navigation display shows the aircraft's track over the ground. The track, which is the result of aircraft heading and winds, tells you which direction the aircraft is actually flying. Winds make it likely that the track and heading will be different. You can get a very good sense of what the winds are doing by comparing the track and heading of the aircraft.

If the aircraft is flying a heading of 090 degrees and the track is 080 degrees, the winds are coming from the south. Notice that having a track indication makes it easy to maintain the desired track. To follow the 049-degree desired track to TRACY upon leaving SUNOL, simply fly the heading that results in a track of 049 degrees. The track display eliminates the traditional method of "bracketing" to find a heading that lets you fly the desired track.

Groundspeed and ETA

The display in *Figure 3-17* also shows groundspeed. Again, the navigation display eliminates the need to calculate groundspeed using distance and time. Based on groundspeed and distance from the active waypoint, the navigation page also provides an estimated time of arrival at the active waypoint.

Fuel Used and Time Remaining

Many advanced avionics navigation units offer fuel calculations and fuel state monitoring. Some units automatically load the initial fuel load, while many require the pilot to correctly enter the amount of fuel into the unit as the beginning fuel on board. Some can have installed transducers (sensors) to measure the fuel used, and display fuel used and time remaining at the current consumption rate. Some lower cost units indicate computed fuel consumption values based on fuel burn rates entered by the pilot. This produces an estimate of fuel used and fuel remaining. This estimate is only as accurate as the values entered by the pilot for fuel on board and the consumption rate. Since the pilot is often using the AFM chart data, there is potential for interpretation error. Then, there is the variation error from the factory charts to the specific aircraft being flown. These factors all tend to degrade the accuracy of the fuel calculation based solely on pilot entered data. Other factors such as a fuel burn that is higher than normal, leaks, or other problems are not displayed unless the system actually registers and senses fuel tank real-time status. These errors can affect both types of systems. The pilot must determine what equipment is installed.

Arriving at the Active Waypoint

As the aircraft reaches the active waypoint, there are four new tasks for the pilot: (1) recognizing imminent arrival at the active waypoint; (2) leading the turn to avoid overshooting the course to the next waypoint; (3) making the next waypoint the new active waypoint; and (4) selecting the desired course to the new active waypoint.

All FMS/RNAV computers offer a sequencing mode that greatly simplifies the performance of the first three of these tasks. Sequencing mode provides three services: waypoint alerting, turn anticipation, and waypoint sequencing.

Waypoint Alerting

The first service performed by the sequencing mode is waypoint alerting. Just prior to reaching each active waypoint, waypoint alerting advises the pilot of imminent arrival at the active waypoint. Waypoint alerting is illustrated in *Figure 3-18*.

Figure 3-18. *Waypoint alerting and turn anticipation.*

Turn Anticipation

The second service performed by the sequencing mode is turn anticipation. During waypoint alerting and prior to reaching the active waypoint, the FMS indicates that it is time to begin the turn to fly the desired track to the new active waypoint. The timing of turn anticipation is based on the aircraft's observed groundspeed and the angle of the turn required to track to the next waypoint. If a standard rate turn is begun when the waypoint alerting indication is presented, the pilot should roll out on course when the aircraft reaches the center of the desired track to the new active waypoint. Turn anticipation is also illustrated in *Figure 3-18*.

When turn anticipation is used, the aircraft does not fly directly over the active waypoint. Rather, the computer commands a turn that "rounds the corner" to some degree, giving priority to having the aircraft roll out on the new desired track to the new active waypoint. This function is illustrated in the upper illustration in *Figure 3-19*.

Turn anticipation occurs only when the active waypoint is designated as a fly-by waypoint. A fly-by waypoint is one for which the computer uses a less stringent standard for determining when the aircraft has reached it. By contrast, some waypoints are designated as flyover waypoints. The FMS will not use turn anticipation for a fly-over waypoint; instead, the navigation will lead the aircraft directly over the waypoint (hence the name). A missed approach waypoint is a typical example of a fly-over waypoint. A fly-over waypoint is illustrated in *Figure 3-19*.

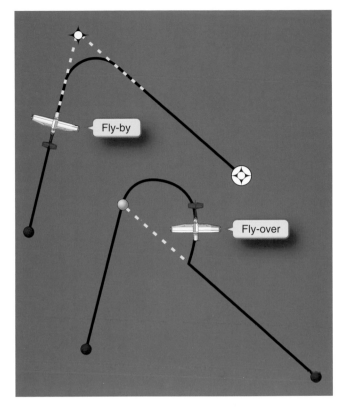

Figure 3-19. *Fly-by and fly-over waypoints.*

Waypoint Sequencing

The third service performed by sequencing mode is waypoint sequencing. Once the aircraft reaches the active waypoint, the FMS automatically makes the next waypoint in the flight plan sequence the new active waypoint. Waypoint sequencing is illustrated in *Figure 3-20*.

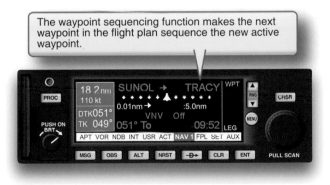

Figure 3-20. *Waypoint sequencing.*

Awareness: Making Waypoint Callouts

To help you stay in touch with the progress of the flight while the FMS automatically performs the navigation task, it is a good practice to announce your arrival (mentally, single pilot; or orally, to the flight crew) at each waypoint in the programmed route. For example, when arriving at SUNOL intersection, you might announce, "Arriving at SUNOL.

TRACY is next. The course is 051 degrees, and the ETE is 10 minutes."

Setting the Course to New Active Waypoint

The last step required when arriving at the active waypoint is to set the course to the next waypoint in the planned route. A PFD such as the one shown in *Figure 3-16* automatically sets the new course on the navigation indicator when the RNAV computer is engaged in sequencing mode. When an FMS is combined with a traditional course deviation indicator, the pilot must manually set the new course using the OBS knob, unless it is an EHSI, or is slaved. "Slaved" means that there is a servo mechanism in the instrument that will respond to the navigation unit.

En Route Sensitivity

When operating en route, the FMS maintains a sensitivity of 5 nautical miles (NM); that is, a CDI displaying course indications from the FMS deflects full-scale when the aircraft drifts 5 NM to either side of the desired track to the active waypoint. An aircraft is considered to be en route when it is more than 30 NM from the origin and destination airports programmed into the flight plan. There are and have been some units that use different values. Consult your specific unit's documentation.

GPS Signal Status

The FMS/RNAV provides position, track, and groundspeed information using signals received from a collection of satellites that are in constant orbit around the Earth when using the GPS navigation source. Although the GPS is highly reliable, satellite reception is sometimes interrupted. Consequently, the pilot must ensure at all times that the system is operational and that it is receiving adequate GPS signals. To simplify this process, all GPS receivers approved for IFR navigation have an automated feature that continually checks the status of the GPS signals. This function is called receiver autonomous integrity monitoring (RAIM). RAIM, which requires adequate simultaneous reception of at least five GPS satellites for IFR navigation reliability, works independently and notifies the pilot when there is a problem with GPS signal reception. When reception problems arise, the FMS/RNAV provides an alert message that notifies the pilot of a GPS reception problem and states that the aircraft position information can no longer be considered reliable. For this reason, regulations require aircraft that are equipped with an RNAV unit using GPS to have an alternate (non-GPS) means of IFR navigation on board (e.g., a VOR receiver) unless the GPS receiver complies with the requirements of WAAS TSO-146B.

For all of the previously mentioned reasons, many manufacturers have coupled inertial navigation units with the FMS to deliver unequalled reliability in navigation. Many more complex FMS units also search or scan for distance measuring equipment (DME) signals and VOR signals as additional navigational sources to compute a "blended" position, which is the best calculation from all of the sources and how the FMS is programmed to "consider" the signals for accuracy. GPS RNAV units usually use only GPS signal sources, but may be able to receive VOR and DME signals as well. In the future, many GPS units will probably receive eLORAN as well, since it is a long range navigation system with greatly improved accuracy as compared to the older LORAN-C. One advantage of the LORAN ("e" or "C") system is that it is ground based and can be easily maintained, as compared to space-based navigation sources.

Accessing Navigational Information En Route

One of the most useful features of an FMS database is its ability to provide quick access to navigational information. Most units allow the pilot to access information about airports, navigation facilities, airway intersections, and other kinds of waypoints. *Figure 3-21* illustrates how one FMS is used to access radio frequencies for an airport.

Figure 3-21. *Accessing communications frequencies in the FMS.*

Essential Skills

1. Select and monitor the en route portion of the programmed flight route, determining waypoint arrival, approving turn anticipation, and waypoint sequencing.

2. Approve or select the correct course automatically displayed or manually tuned.

3. Determine if FMS makes fuel calculations and what sensors and data entries are required to be made by the pilot.

4. Ensure that the track flown is that cleared by air traffic control (ATC).

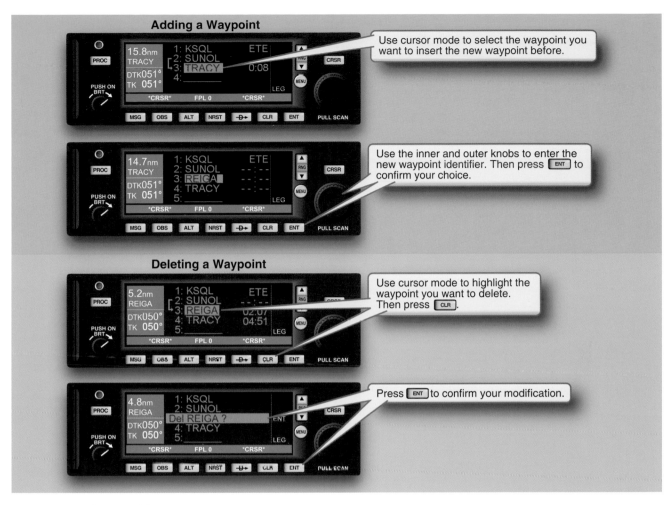

Figure 3-22. *Adding and deleting waypoints.*

5. Determine that the display CDI sensitivity is satisfactory for the segment being flown.

En Route Modifications

Part of the challenge of using FMS en route is dealing with modifications to the planned flight route. This section describes five en route modifications.

Adding and Deleting Waypoints From the Programmed Route

All FMS/RNAV units allow en route (not published departure, arrival or approach) waypoints to be added and deleted to the programmed route. These techniques are illustrated in *Figure 3-22.*

ATC may issue instructions to a point defined by a VOR radial and DME value. The pilot must know how to enter such a waypoint as a user waypoint, name it, and recall it. If the unit's memory is very limited, the pilot should also be adept at removing the waypoint.

Direct To

Another simple modification is one that requires the pilot to proceed directly to a waypoint. In some cases, the waypoint to fly directly toward is one that already appears in the programmed flight plan. In this case, the pilot simply selects that waypoint in the flight plan and activates the direct-to function, as illustrated in *Figure 3-23.*

The direct-to waypoint now becomes the active waypoint. After reaching this waypoint, the system proceeds to the next waypoint in the programmed route.

In other cases, you may be asked to fly directly to a waypoint that does not already appear in the programmed flight route. In this case, one strategy is to add the waypoint to the programmed route using the technique illustrated in *Figure 3-22,* and then proceed directly to the waypoint using the technique illustrated in *Figure 3-23.* Another option is to use the direct-to function to get the flight started toward the assigned waypoint, and then add the new waypoint to the appropriate place in the programmed flight plan.

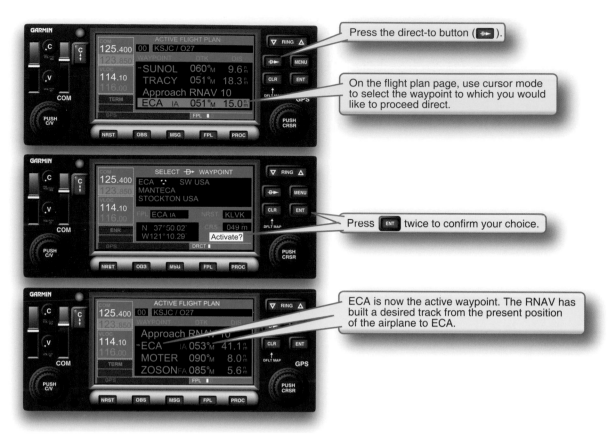

Figure 3-23. *Direct-to operation.*

Risk: What Lies Ahead on a Direct-To Route?

The direct-to function offers a convenient way to shorten your time and distance en route if ATC authorizes that track. When you perform a direct-to operation, though, remember that the system builds a new track from your present position to the new waypoint. This track does not necessarily correspond to any previously planned airway or route, so it is critical to ensure that your new direct route is clear of all significant obstructions, terrain, weather, and airspace.

Cancel Direct To

ATC may sometimes cancel a previously issued direct-to clearance and ask you to resume the previously cleared route. Most FMSs offer a simple way of canceling a direct-to operation. *Figure 3-24* illustrates the procedure for one FMS.

Selecting a Different Instrument Procedure or Transition

ATC will sometimes issue an instrument procedure or transition that is different from what you would expect. Entering a new procedure or transition is usually a simple matter of making new menu choices, as illustrated in *Figure 3-25*. In most units, if you are training or wish to fly the approach again, you must learn how to set the selector or cursor back to the initial fix, which will restart the approach sequence.

Figure 3-24. *Canceling a direct-to operation.*

Proceeding Directly to the Nearest Airport

One of the most useful features of an FMS is its ability to provide you with immediate access to a large navigation database. This feature is particularly useful when a suitable nearby airport or navigation facility must be located quickly. *Figure 3-26* shows how to locate and proceed directly to the nearest suitable airport using one manufacturer's system.

Essential Skills

1. Proceed directly to a waypoint in the programmed route.

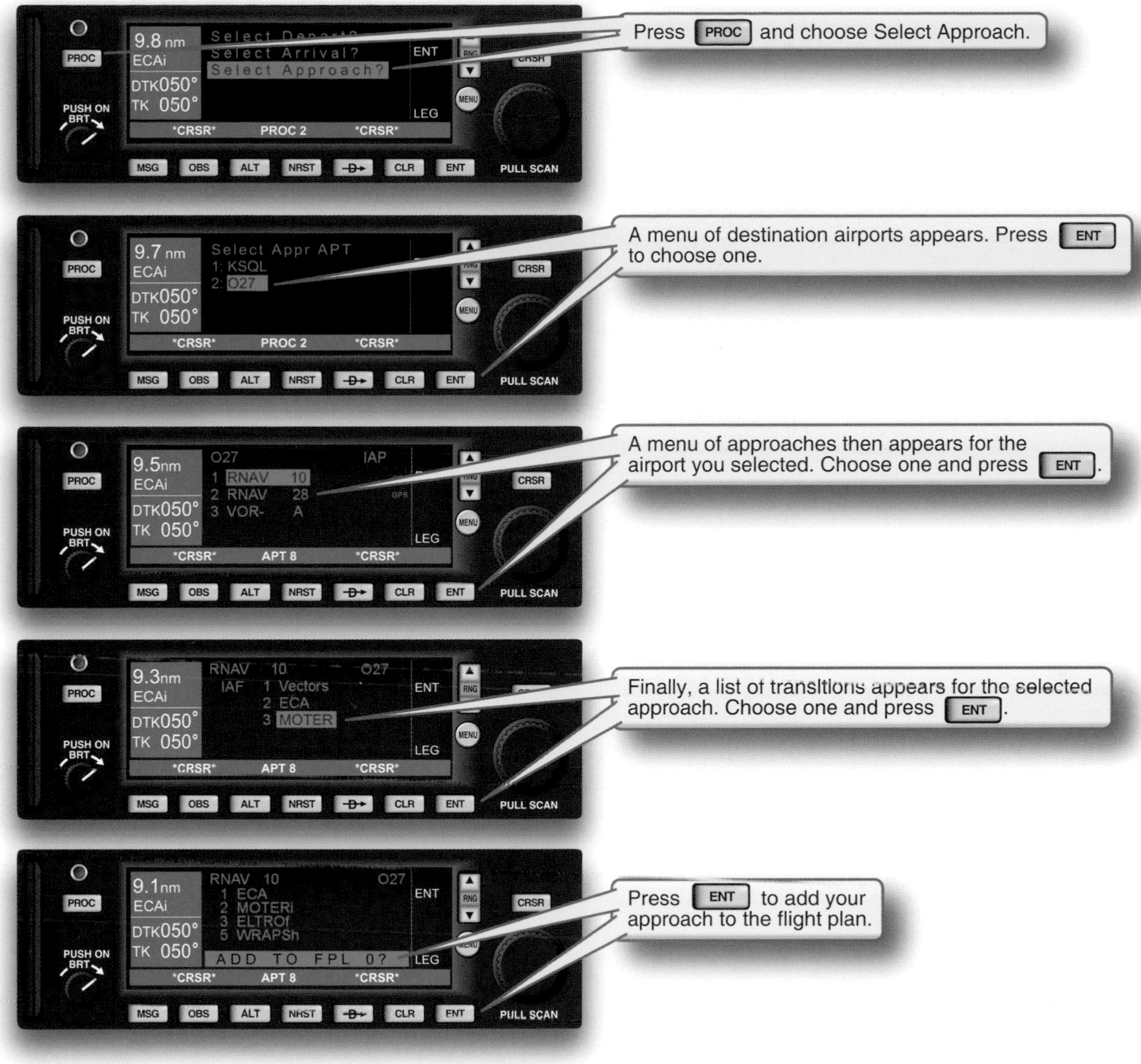

Figure 3-25. *Selecting a different instrument procedure or transition.*

Press **PROC** and choose Select Approach.

A menu of destination airports appears. Press **ENT** to choose one.

A menu of approaches then appears for the airport you selected. Choose one and press **ENT**.

Finally, a list of transitions appears for the selected approach. Choose one and press **ENT**.

Press **ENT** to add your approach to the flight plan.

2. Cancel a programmed or selected waypoint or fix.

3. Select a different instrument procedure or transition.

4. Restart an approach sequence.

5. Immediately find the nearest airport or facility.

6. Edit a flight plan.

7. Enter a user waypoint.

Descent

Making the transition from cruise flight to the beginning of an instrument approach procedure sometimes requires arriving at a given waypoint at an assigned altitude. When this requirement is prescribed by a published arrival procedure or issued by ATC, it is called a crossing restriction. Even when ATC allows a descent at the pilot's discretion, you need to choose a waypoint and altitude for positioning convenient

Highlight any airport in cursor mode and press ⊕ to receive guidance directly to that airport.

Use the outer knob to select the nearest page. The first subpage shows the nearest airports.

Figure 3-26. *Proceeding directly to the nearest airport.*

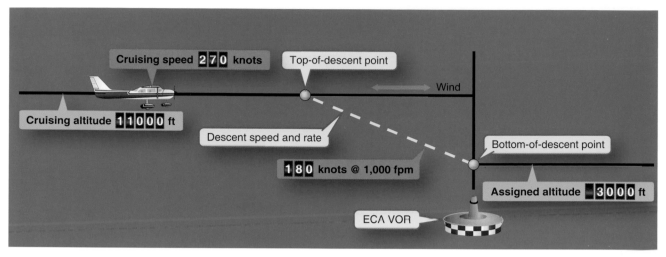

Figure 3-27. *The descent planning task.*

Elements of Descent Planning Calculations

Figure 3-27 illustrates the basic descent planning task. The task begins with an aircraft flying at an assigned cruising altitude. The aircraft must descend to an assigned altitude and reach that assigned altitude at a designated bottom-of-descent point. The next step is to choose a descent rate and a descent speed. The ultimate goal is to calculate a top-of-descent point, which is the point at which, if you begin the descent and maintain the planned descent rate and airspeed, you will reach the assigned altitude at the designated bottom-of-descent point.

In a basic aircraft, you must rely on manual calculations to perform the descent planning task. In an advanced avionics aircraft, there are two descent planning methods available: (1) manual calculations, and (2) the vertical navigation features of the FMS unit. Skillful pilots use both methods and cross-check them against one another in order to reduce the possibility of error and help keep the pilot "in the loop."

Manual Descent Calculations

The simplest technique for calculating the distance required to descend uses a descent ratio. The table in *Figure 3-28* lists a descent ratio for many combinations of planned descent speeds and descent rates. Calculating a descent is a simple matter of looking up the descent ratio for your target descent rate and groundspeed, and multiplying the descent ratio by the number of thousands of feet in altitude that you must descend. For example, suppose you are asked to descend from 11,000 feet to meet a crossing restriction at 3,000 feet. Since there is a 200-knot speed restriction while approaching the destination airport, you choose a descent speed of 190 knots

and a descent rate of 1,000 feet per minute (fpm). Assuming a 10-knot headwind component, groundspeed in the descent is 180 knots. Referring to the table in *Figure 3-28,* the planned descent speed and rate indicate a ratio of 3.0. This means that you will need 3 NM for every 1,000 feet of descent. You must descend a total of 8,000 feet (11,000 feet – 3,000 feet). A total of 24 NM is needed to descend 8,000 feet (3 NM × 8 = 24 NM), and must, therefore, begin the descent 24 NM away from the end-of-descent point.

Another technique for calculating descents is to use the formula shown in *Figure 3-29.* A descent table can be found in the front of each set of U.S. Terminal Procedures on page D-1. Working through the formula for the ECA VOR crossing restriction example, 8 minutes is needed to descend 8,000 feet at the planned descent rate of 1,000 fpm. At your planned descent speed of 180 knots, you will cover 3 NM per minute. Thus, in 8 minutes, you will cover 24 NM. Once again, you must start the descent 24 NM prior to ECA to meet the crossing restriction.

Coordinating Calculations with Aeronautical Charts

Regardless of which method is used, it is always a good idea to locate the top-of-descent point chosen on the aeronautical chart. *Figure 3-30* shows a chart that covers the area surrounding the ECA VOR. A top-of-descent point 24 NM prior to ECA is located 3 NM before PATYY intersection.

Alternate Navigation Planning

Using the aeronautical chart to locate the top-of-descent point has a second advantage. Since regulations require you to have an alternate means of navigation onboard if the computer does not comply with TSO 146B, the aeronautical chart allows you to check minimum altitudes for VOR reception

A descent ratio table is provided for use in planning and executing descent procedures under known or approximate ground speed conditions and rates of descent. The ratio expresses the number of nautical miles needed to descend 1,000 ft.

DESCENT GRADIENT RATE (ft./min)	GROUND SPEED (knots)											
	90	100	120	140	160	180	200	220	240	260	280	300
500	3.0	3.3	3.7	4.6	5.3	6.0	6.7	7.3	8.0	8.7	9.3	10.0
600	2.5	2.8	3.1	3.9	4.4	5.0	5.6	6.1	6.7	7.2	7.8	8.3
700	2.1	2.4	2.6	3.3	3.8	4.3	4.8	5.3	5.7	6.2	6.7	7.1
800	1.9	2.1	2.3	2.9	3.3	3.8	4.2	4.6	5.0	5.4	5.8	6.3
900	1.7	1.9	2.0	2.6	3.0	3.3	3.7	4.1	4.4	4.8	5.2	5.6
1000	1.5	1.7	1.8	2.3	2.7	3.0	3.3	3.7	4.0	4.3	4.7	5.0
1100	1.4	1.5	1.7	2.1	2.4	2.7	3.0	3.3	3.6	3.9	4.2	4.5
1200	1.3	1.4	1.5	1.9	2.2	2.5	2.8	3.1	3.3	3.6	3.9	4.2
1300	1.2	1.3	1.4	1.8	2.1	2.3	2.6	2.8	3.1	3.3	3.6	3.8
1400	1.1	1.2	1.3	1.7	1.9	2.1	2.4	2.6	2.9	3.1	3.3	3.6
1500	1.0	1.1	1.2	1.6	1.8	2.0	2.2	2.4	2.7	2.9	3.1	3.3
1600	0.9	1.0	1.1	1.5	1.7	1.9	2.1	2.3	2.5	2.7	2.9	3.1
1700	0.9	1.0	1.1	1.4	1.6	1.8	2.0	2.2	2.4	2.5	2.7	2.9
1800	0.8	0.9	1.0	1.3	1.5	1.7	1.9	2.0	2.2	2.4	2.6	2.8
1900	0.8	0.9	1.0	1.2	1.4	1.6	1.8	1.9	2.1	2.3	2.5	2.6
2000	0.7	0.8	0.9	1.2	1.3	1.5	1.7	1.8	2.0	2.2	2.3	2.5

Figure 3-28. *Descent ratio table.*

$$\frac{\text{Cruising Altitude (ft)} - \text{Descent Altitude (ft)}}{\text{Descent Rate (ft/min)}} \times \frac{\text{Groundspeed (NM/hr)}}{60 \text{ (min/hr)}} = \text{NM required}$$

$$\frac{11,000 \text{ ft} - 3,000 \text{ ft}}{1,000 \text{ ft/min}} \times \frac{180 \text{ NM/hr}}{60 \text{ (min/hr)}} - \frac{8,000 \text{ ft}}{1,000 \text{ ft/min}} \times 3 \text{ NM/min} = 8 \times 3 \text{ NM} = 24 \text{ NM}$$

Figure 3-29. *Descent formula.*

along the route of flight in case VOR navigation is required at any time. The airway that leads to the ECA VOR lists a minimum en route altitude (MEA) of 3,000 feet, which is the clearance altitude.

Calculating Descents with the FMS

Building a descent with an FMS follows the familiar process of entering the basics of the descent into the system, letting the system do the math, and then reviewing what the system has produced. Most FMS units offer a descent planning or vertical navigation (VNAV) page that allows you to enter the details of your descent. *Figure 3-31* shows the VNAV page for one manufacturer's system. Note that there is an entry for each of the descent planning concepts discussed above. Computers perform the calculations using the same formulas and data.

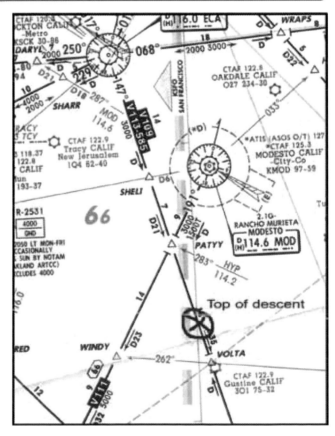

Figure 3-30. *Top-of-descent point on an en route chart.*

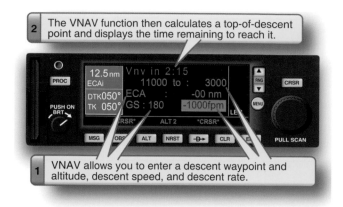

2 The VNAV function then calculates a top-of-descent point and displays the time remaining to reach it.

1 VNAV allows you to enter a descent waypoint and altitude, descent speed, and descent rate.

Figure 3-31. *Planning a descent with an advanced avionics unit.*

It is a good idea to cross-check the results of your manual descent calculations with the results produced by the computer. Many RNAV units do not display a waypoint for the planned top-of-descent point. However, there may be an "approaching VNAV profile" message that anticipates the descent point and cues the pilot to begin descending. Caution is advised that some systems calculate the vertical flight path dependent on the current airspeed/groundspeed values. Lowering the nose and gaining airspeed in the descent may confuse you into perceiving a false vertical goal or vertical rate, resulting in failure to meet the crossing restriction with some systems. Determine if the system recomputes the airspeed/groundspeed, or if you must enter the descent airspeed during the VNAV programming.

Managing Speed

Up to this point the focus has been on the task of losing excess altitude. For example, in the situation shown in *Figure 3-27,* you are faced with the requirement to reduce altitude from 11,000 feet to 3,000 feet. Most descent scenarios also present the challenge of losing excess speed. In piston aircraft of modest performance, losing excess speed seldom

requires much forethought. Slowing from a cruising speed of 120 knots to an approach speed of 100 knots requires little planning and can be accomplished quickly at almost any point during a descent. Flying higher performance aircraft requires a closer look at concepts of excess altitude and excess speed. Higher performance piston engines usually require descent scheduling to prevent engine shock cooling. Either the engines must be cooled gradually before descent, or power must be constant and considerable in the descent to prevent excessive cooling. In such instances, a much longer deceleration and gradual engine cooling must be planned to prevent powerplant damage. Additionally, the turbulence penetration or V_A speeds should be considered with respect to weather conditions to avoid high speeds in turbulent conditions, which could result in overstressing the airframe. Drag devices such as spoilers can be of great advantage for such maneuvers. In the scenario in *Figure 3-27,* a cruising speed of 270 knots is inappropriate as the aircraft descends below 10,000 feet, and even more so as it enters Class C airspace. Therefore, descent planning must include provisions for losing excess airspeed to meet these speed restrictions.

Some sophisticated FMSs are able to build in a deceleration segment that can allow the aircraft to slow from the cruise speed to the desired end-of-descent speed during the descent. This type of navigation system allows you to maintain the cruise speed up until the top-of-descent point and calculates the deceleration simultaneously with the descent. A deceleration segment is illustrated in *Figure 3-32.*

Simple FMS units such as GPS RNAV receivers assume that you will slow the aircraft to the planned descent speed before reaching the top-of-descent point. ATC timing may preclude this plan.

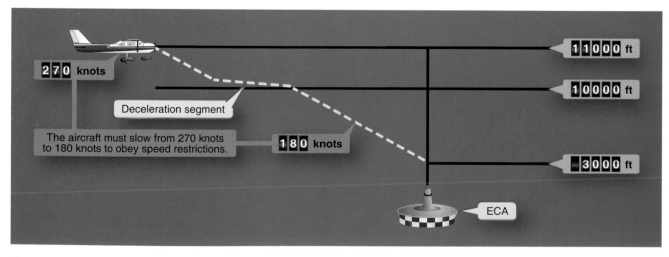

Figure 3-32. *A deceleration segment planned by a more sophisticated FMS.*

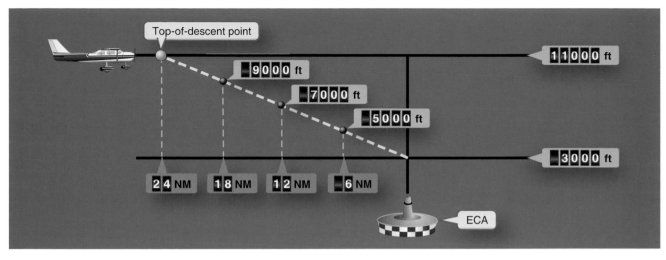

Figure 3-33. *Planned descent path as a wire in the sky.*

Descent Flying Concepts

Probably the most important descent flying concept to understand is that a planned descent is basically a "pathway in the sky," similar to the glideslope associated with an ILS procedure. If you start down at the planned top-of-descent point, fly a groundspeed of 180 knots, and descend at 1,000 feet per minute (fpm), you will be flying on a fixed path between the top-of-descent point and the bottom-of-descent point. If you maintain the 180-knot and 1,000-foot-per-minute descent, you will cross a point 18 NM from ECA at exactly 9,000 feet, a point 12 NM from ECA at 7,000 feet, and a point 6 NM from ECA at exactly 5,000 feet, as shown in *Figure 3-33*.

If you are at a different altitude at any of these points, you will not cross ECA at the required 3,000 feet unless corrective action is taken. Four things can cause you to drift from a planned descent path:

1. Not following the planned descent rate
2. Not following the planned descent speed
3. Unexpected winds
4. Navigation system not recalculating for airspeed change of descent

Figure 3-34 shows the effect of each situation on the position of the aircraft with respect to the planned descent path.

Flying the Descent

The key to flying a descent is to know your position relative to the pathway-in-the-sky at all times. If you drift off the path, you need to modify the descent speed and/or descent rate in order to rejoin the descent path. Many FMSs do not give a direct indication of progress during a descent. You must be very familiar with the indirect indications of the VNAV descent. In this case, follow the planned descent rate

and speed as closely as possible and be mindful of altitude and position while approaching the crossing restriction fix.

Determining Arrival at the Top-of-Descent Point

All navigation systems provide some type of alert informing the pilot of arrival at the planned top of descent point, and that it is time to begin the descent at the speed and rate entered into the FMS.

If air traffic control is able to accommodate your request, the ideal point to begin the descent is at the planned top-of-descent point. If air traffic control is unable to accommodate such a request, one of two scenarios will ensue: an early descent or a late descent.

Early Descents

Beginning descent before reaching the planned top-of-descent point means you must set aside descent planning and proceed without the benefit of vertical guidance offered by the navigation system. If, during the descent, the navigation computer does not display position with respect to the planned descent path, you must simply do the best possible to arrive at the crossing restriction at the assigned altitude. If the navigation system does display position with respect to the planned descent path, you can usually recapture the planned descent path and resume flying with vertical guidance from the computer. The basic technique is to initiate descent at a reasonable descent rate that is less than the planned descent rate. If you follow this initial descent rate, you will eventually intercept the planned descent path, as shown in *Figure 3-35*.

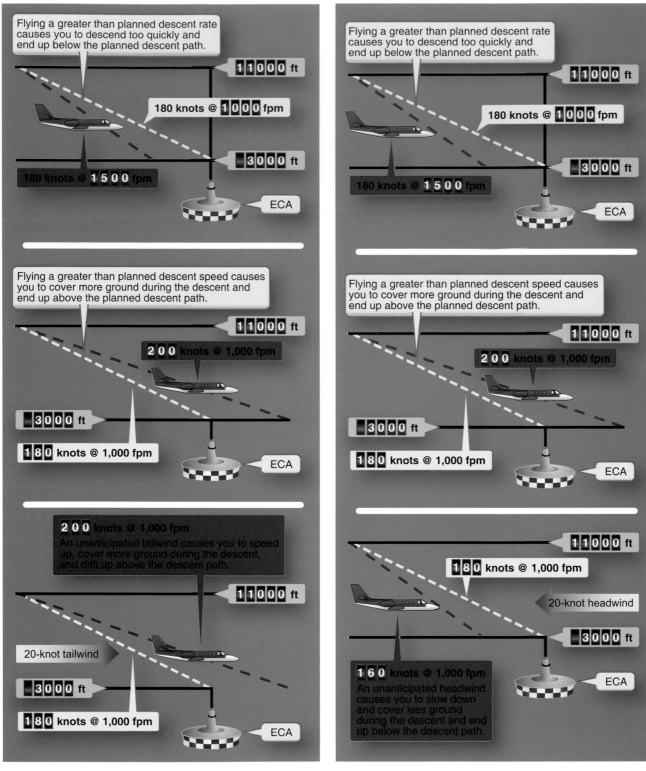

Figure 3-34. *Drifting off the planned descent path.*

Figure 3-35. *Early descent scenario.*

Late Descents

Beginning the descent beyond the planned top-of-descent point means that you will have the same amount of excess altitude, but a shorter distance and time to lose it, as shown in *Figure 3-36.*

Since flying beyond the planned top-of-descent point leaves less time to lose excess altitude, your goal is to minimize the "overrun" distance by slowing the aircraft as soon as a late descent scenario is suspected. A lower speed means you will

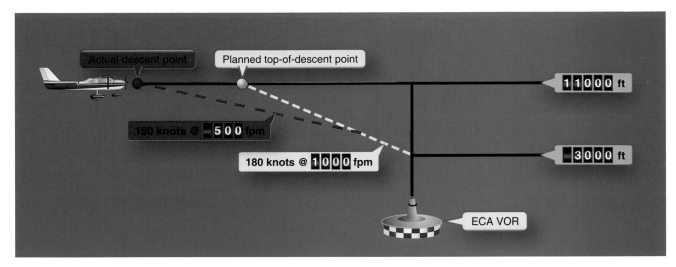

Figure 3-36. *Late descent scenario.*

cover less distance in the same amount of time, and thus be left with more time to lose altitude.

Common Error: Not Considering Winds During Descent Planning

A common error in planning a descent is failing to consider winds and their effect on groundspeed. As illustrated in *Figure 3-34,* if you fail to take into account a 20-knot tailwind, your groundspeed will be faster than you planned, and you will reach the target waypoint before reaching the assigned altitude.

Essential Skills

1. Determine the descent airspeed to be used with attention to turbulence, aircraft descent profile, and powerplant cooling restrictions.

2. Program, observe, and monitor the top of descent, descent rate, and level-off altitude.

3. Plan and fly a descent to a crossing restriction.

4. Recognize and correct deviations from a planned descent path, and determine which factor changed.

Intercept And Track Course

Intercepting and Tracking a Different Course to the Active Waypoint

Figure 3-37 illustrates a common situation. Air traffic control instructs you to fly to a waypoint via an inbound course different from the desired track calculated by the FMS. In the example in *Figure 3-37,* you are en route to SUNOL intersection. The FMS has calculated a desired track of 060 degrees, but ATC has instructed you to fly a heading of 080 degrees to intercept a 009-degree course to SUNOL.

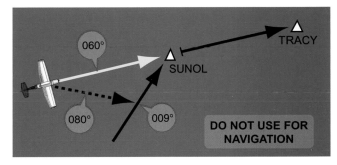

Figure 3-37. *A simple course intercept scenario.*

The FMS is set to take the aircraft to SUNOL intersection, but via an inbound course different from the one ATC has cleared you to follow. Therefore, you need to be a means of programming the FMS to follow your choice of course instead of the desired track that it has identified.

The Nonsequencing Mode

Every IFR-capable FMS/RNAV unit offers an alternative mode of operation, the nonsequencing mode, which allows you to perform this particular task. Like the OBS knob which allows you to select VOR radials, the nonsequencing mode allows you to select courses to or from an active waypoint. The nonsequencing mode differs from the sequencing mode in two important ways:

1. Nonsequencing mode allows you to select a different inbound course to the active waypoint. For this reason, some manufacturers refer to the nonsequencing mode as OBS (hold or suspend) mode, which suggests similarity to the OBS knob found on traditional VOR indicators. As the OBS knob allows you to select inbound VOR radials, the nonsequencing mode allows you to select inbound courses to an active waypoint.

Figure 3-38. *Setting a different course to the active waypoint.*

2. Nonsequencing mode stops the waypoint sequencing feature of the FMS/RNAV unit. If engaged in nonsequencing mode, the FMS/RNAV program does not automatically sequence to the next waypoint in the flight plan when the aircraft arrives at the active waypoint.

Every FMS/RNAV offers a way to switch to the nonsequencing mode. There is typically a button marked OBS (or Hold), and an OBS or course selection knob to select an inbound course to the active waypoint. Figure 3-38 illustrates the procedure for one particular FMS.

Once you switch to nonsequencing mode and select the inbound course of 009°, the navigation indicator reflects aircraft position with respect to the 009° course. The navigation indicator in *Figure 3-38* shows that you are west of course. The assigned heading of 080° provides an acceptable intercept angle. As you fly the 080° heading, the needle centers as you reach the 009° course. Once the 009° course is reached and the needle has centered, you can turn to track the 009° course inbound to SUNOL.

It is important to remember that the nonsequencing mode suspends the FMS/RNAV's waypoint sequencing function. If you reach SUNOL and the unit is still set in the nonsequencing mode, it will not sequence on to the next waypoint. Generally, once established on a direct course to waypoint or navaid, switching back to sequencing (releasing the Hold or Suspend function) mode allows the FMS/RNAV to continue to the programmed point and thence onward according to the programmed routing. Setting the computer back to the sequencing mode is usually accomplished by pressing the OBS (Hold or Suspend) button again.

Common Error: Forgetting To Re-Engage Sequencing Mode After Course Intercept

By far the most common error made with the nonsequencing mode is forgetting to re-engage the sequencing mode once the course has been intercepted. The result is that the FMS will not sequence to the next waypoint in the flight route upon reaching the active waypoint. The best indicator of this event is the "To/From" navigation display showing "From." Normally all FMS fly "To" the waypoint, unless that unit does holding patterns. Flying "From" a waypoint can only be done in the "OBS"/"Hold"/"Suspend" mode.

Awareness: Remembering To Make Needed Mode Changes

The use of the sequencing and nonsequencing modes illustrates another aspect of maintaining good mode awareness—remembering to make required mode changes at future times during the flight. Remembering to do tasks planned for the future is a particularly error-prone process for human beings. Aviation's first line of defense against such errors is the checklist. Creating your own checklist or callout procedures for maneuvers such as course intercepts is a good way to minimize this error. For example, a simple callout procedure for the course intercept maneuver might commence when the aircraft nears the point of interception. "Course is alive. Course is captured. Changing back to sequencing mode."

Intercepting and Tracking a Course to a Different Waypoint

Figure 3-39 illustrates a slightly more complicated request often made by air traffic control. While en route to SUNOL, ATC instructs you to fly a heading of 060° to intercept and track the 049 course to TRACY. This situation requires two separate tasks: changing not only the inbound course, but also the active waypoint.

The first step is to change the active waypoint using the direct-to function, as illustrated in *Figure 3-40*. Remember, though, that if you use the direct-to function to make TRACY the active waypoint, the FMS calculates a desired track that takes you from the present position to TRACY intersection.

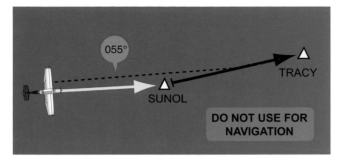

Figure 3-39. *A more complicated course intercept scenario.*

The second step, illustrated in *Figure 3-40*, is to change the desired track to TRACY by setting the computer in the nonsequencing mode and selecting the inbound course. You now continue on the assigned heading until the needle centers, then set the FMS back to the sequencing mode, and continue inbound on the assigned course to TRACY intersection.

Common Error: Setting the Wrong Inbound Course During a Course Intercept

One common error made during course intercepts is to select the wrong course to the active waypoint. Some FMSs automatically set the course indicator ("slew" the needle) to the inbound course. Where this capability does not exist, pilots occasionally select the heading that they have been assigned to fly to intercept the course instead of the inbound course. The outcome of this error is illustrated in *Figure 3-41*.

Common Error: Setting the Wrong Active Waypoint During a Course Intercept

Another common error is failing to realize that ATC has instructed you to intercept a course to a different waypoint. *Figure 3-42* shows the outcome when the pilot neglects to set TRACY as the active waypoint in the previous example. The FMS offers guidance along the correct course, but to the wrong waypoint.

Catching Errors: A Helpful Callout Procedure for Course Intercepts

The following is a useful technique for avoiding two errors commonly made during course intercept maneuvers. Ask yourself the following two questions when working your way through any course intercept maneuver:

Question #1: Where am I going?

Point to the active waypoint on the navigation page and make sure it shows the waypoint that you wish to fly toward.

Question #2: How am I getting there?

Point out the desired track to the active waypoint on the navigation page. If it is not the one you want, engage the nonsequencing mode and select the course you want.

Figure 3-40. *Intercepting a course to a different waypoint.*

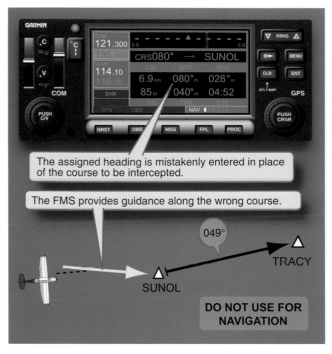

Figure 3-41. *Selecting the wrong course to the active waypoint.*

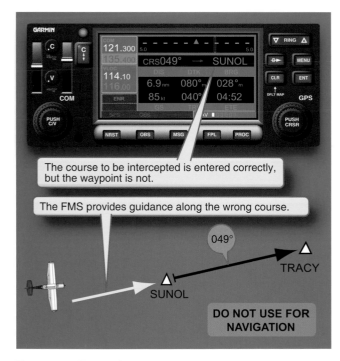

Figure 3-42. *Setting the wrong active waypoint.*

Essential Skills

1. Program and select a different course to the active waypoint.

2. Select the nonsequencing waypoint function (OBS, Hold, or Suspend) to select a specified navigation point.

3. Reactivate the sequencing function for route navigation.

Holding

The FMS/GPS unit's nonsequencing mode provides an easy way to accomplish holding procedures. When instructed to hold at a waypoint that appears in the route programmed in the FMS/GPS unit, simply engage the nonsequencing mode prior to reaching the waypoint. With waypoint sequencing suspended, you can determine and fly the appropriate holding pattern entry, select the inbound holding course using the course selector (OBS in some) knob or buttons, and fly the holding pattern while timing the outbound leg. Some FMSs can automatically enter the holding pattern, and continue to hold if programmed. As the aircraft repeatedly crosses the holding waypoint with each turn in the hold, the holding waypoint remains the active waypoint. When you are cleared out of the holding pattern for the approach or to another point, you should select the sequencing mode or cancel the suspension before reaching the holding waypoint for the last time. When you pass the holding waypoint in sequencing mode, the FMS/GPS unit will then sequence to the next waypoint in the route. This procedure is demonstrated in *Figure 3-43*.

Preprogrammed Holding Patterns

Some FMS/GPS units insert preprogrammed holding patterns into published instrument procedures. The purpose of a preprogrammed holding pattern is to relieve you of many of the tasks described above for flying a holding pattern.

Figure 3-44 illustrates a preprogrammed holding pattern that appears at the end of a missed approach procedure.

As the FMS/GPS unit shown in *Figure 3-44* sequences to a preprogrammed holding pattern, the navigation displays a message indicating the type of holding entry required based on the aircraft's current track. The system then automatically switches to a special nonsequencing mode that not only stops waypoint sequencing, but also sets the inbound course to the holding waypoint. This special nonsequencing mode is different from the nonsequencing mode you engage manually. In *Figure 3-44,* this system uses the term suspend mode (SUSP) to signify the nonsequencing mode that is automatically engaged during a preprogrammed holding procedure. Depending on the type of holding procedure, the unit may or may not automatically switch back to the sequencing mode after the aircraft crosses the holding fix. As always, you must be careful to maintain constant mode awareness.

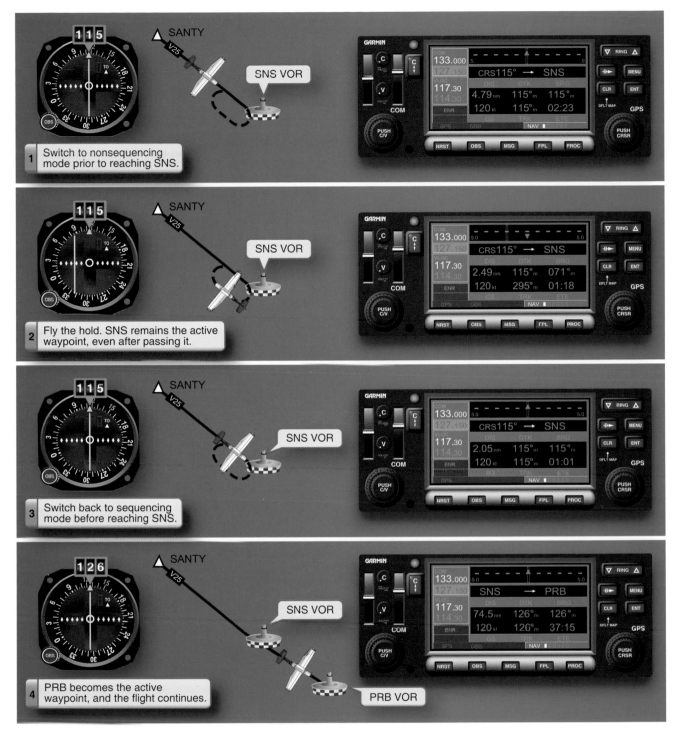

Figure 3-43. *Using the nonsequencing mode to fly a hold.*

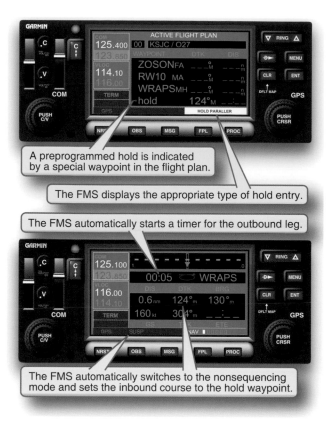

A preprogrammed hold is indicated by a special waypoint in the flight plan.

The FMS displays the appropriate type of hold entry.

The FMS automatically starts a timer for the outbound leg.

The FMS automatically switches to the nonsequencing mode and sets the inbound course to the hold waypoint.

Figure 3-44. *A preprogrammed holding procedure.*

Common Error: Mismanaging the Sequencing/Nonsequencing Modes During a Hold

Mismanagement of the sequencing and nonsequencing modes during a holding procedure is another common error. Failing to switch the FMS/GPS RNAV to the nonsequencing mode prior to reaching the holding waypoint, or prematurely switching the unit back to the sequencing mode once established in the hold, can prompt the FMS/GPS to sequence past the holding waypoint. In this case, you are left without guidance along the inbound holding course.

Essential Skills

1. Select a preprogrammed holding pattern, or nonsequencing mode.

2. Select and setup a non-preprogrammed holding pattern inbound course.

3. Determine the proper sequence of software commands for the holding pattern, transition to approach, approach, and MAP navigation.

ARCS

FMS and some GPS units simplify the problem of tracking arcs, which are curved courses between waypoints. The key feature of an arc is that there is no one bearing that takes you from one waypoint to the next. Rather, depending

on its length, an arc requires you to follow a gradually changing heading toward the active waypoint. The example in *Figure 3-45* illustrates how an FMS is used to fly a DME arc procedure.

Essential Skills

1. Select an approach procedure with an arc.

2. Select the course, or determine that automatic course CDI setting will occur.

GPS and RNAV (GPS) Approaches

An IFR-capable GPS RNAV/FMS with qualified GPS receiver(s) can be used as the sole means of navigation for several kinds of instrument approach procedures, but you need to know which approaches can be used with your particular GPS RNAV unit. The following paragraphs review the approaches available today.

A GPS overlay approach is illustrated in *Figure 3-46*. The basic benefit of the GPS overlay approach is that it allows use of an IFR approved GPS receiver to navigate and fly a conventional nonprecision approach. From the previous text, you must know how to hold the specific sequences and how the unit can be stopped from sequencing through the flight plan. Many approaches require holding or a procedure turn to orient the aircraft correctly for the approach course. If you cannot control the sequencing of the FMS, you will lose course guidance upon the turn for outbound holding, as the FMS/GPS receiver sequences for the course beyond the holding fix.

GPS overlay approaches are named for the conventional system upon which the approach is based, but include the word GPS. The approach in *Figure 3-46* is based on an existing NDB approach. If the aircraft has an IFR-approved FMS/GPS RNAV, you may use that guidance to fly the GPS overlay approach. It is not necessary for the aircraft to have the conventional navigational equipment on board for that approach, but conventional navigational avionics will be required for any required alternate, if equipped with a TSO-129 GPS receiver. If conventional avionics are installed in the aircraft, there is no requirement to use the equipment in any way, although monitoring is always a good practice. If the installed FMS/GPS receiver is TSO-145A/146A WAAS certified, no other navigation equipment is required.

One common pitfall of all advanced avionics approaches is the sometimes limited notification of the position along the approach path. In many instances, you must read the name of the waypoint to confirm where the aircraft is headed. It is easy for you to be preoccupied with cross-check and flying

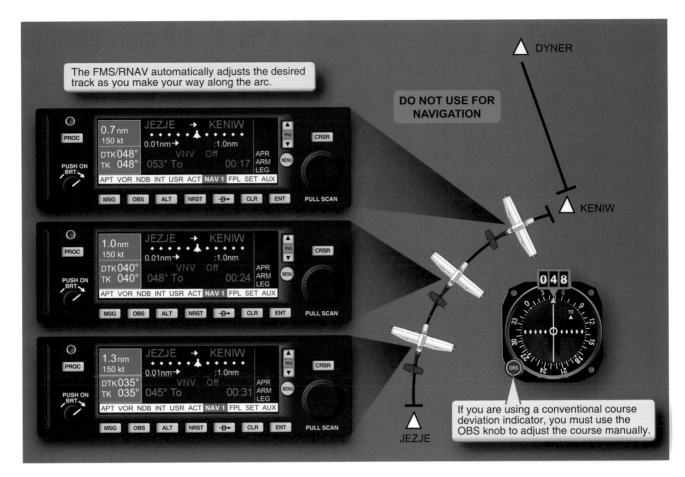

Figure 3-45. *Flying an approach with an arc.*

duties to miss a waypoint change and be of the mindset that you have one more waypoint to go before descent, or even worse, before a missed approach. Two main values always to include in the cross-check are:

1. Verification of the waypoint flying "to."

2. Verification that the distance to the waypoint is decreasing. Upon reaching the missed approach point (MAP), the system will automatically go to "Suspend," "Hold," or "OBS" at the MAP, and the distance to go will begin counting up or increasing as the distance from the MAP behind increases. Acknowledge the MAP and the beginning of the MAP segment by an action (button, knob, etc.) to allow sequencing to the holding point or procedure.

Not all units delay commanding a turn prior to reaching the specified turn altitude. You must know the required navigation courses and altitudes. The FMS/GPS unit may not be 100 percent correct, especially if an ADC is not installed.

Since the FMS/GPS automatically switches to the approach sensitivity, you must not attempt to use the "approach" mode of the autopilot at that time, unless the autopilot documentation specifically directs the use of that mode at that time. Using that mode would make the autopilot hypersensitive and too responsive to navigation signals.

GPS stand-alone approaches are nonprecision approaches based solely on the use of the GPS and an IFR-capable FMS with GPS navigation receiver or GPS RNAV. A GPS stand-alone approach is shown in *Figure 3-47.*

RNAV (GPS) approaches are designed to accommodate aircraft equipped with a wide variety of GPS receivers. An RNAV (GPS) approach procedure is shown in *Figure 3-48.* A GPS approach typically offers different approach minimums (and sometimes different missed approach points) depending on the type of GPS receiver, aircraft, and installation being used to complete the approach.

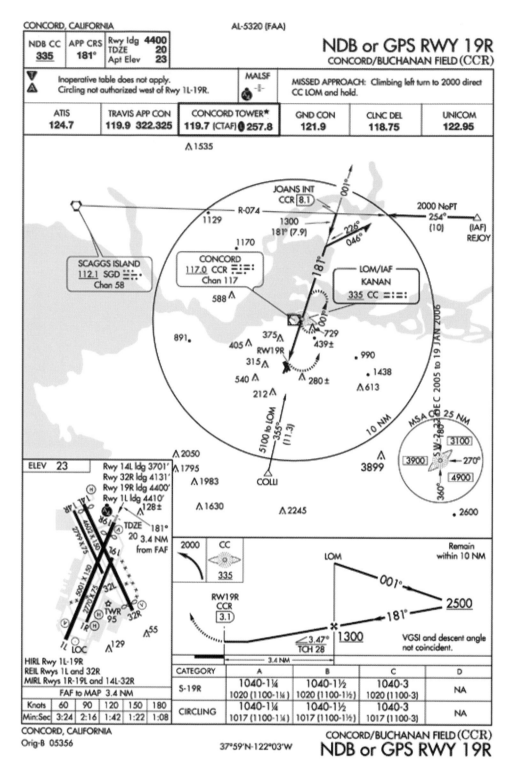

NOT FOR USE IN NAVIGATION

Figure 3-46. *A GPS overlay approach.*

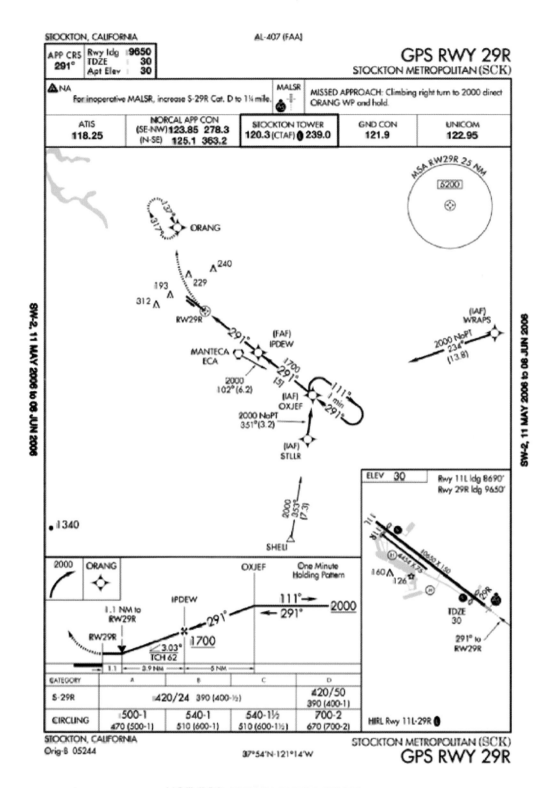

NOT FOR USE IN NAVIGATION

Figure 3-47. *GPS stand-alone approach.*

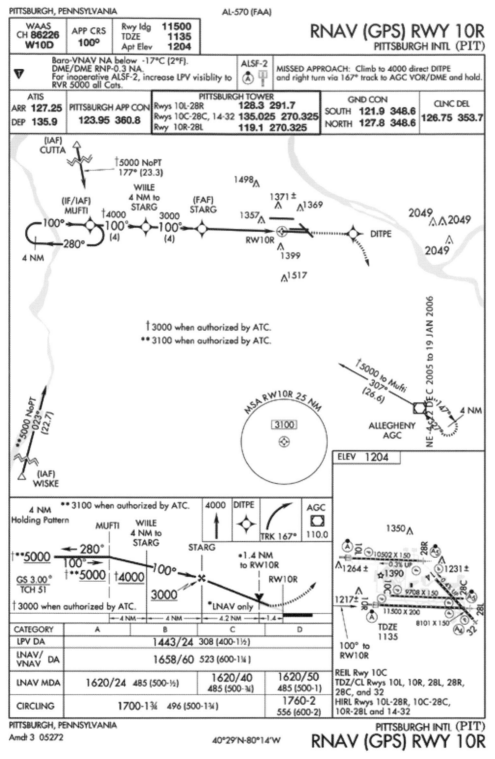

Figure 3-48. *RNAV (GPS) approach.*

LNAV

LNAV (lateral navigation), like a conventional localizer, provides lateral approach course guidance. LNAV minimums permit descent to a prescribed minimum descent altitude (MDA). The LNAV procedure shown on the chart in *Figure 3-48* offers an MDA of 1,620 feet.

LNAV/VNAV

LNAV/VNAV (lateral navigation/vertical navigation) equipment is similar to ILS in that it provides both lateral and vertical approach course guidance. Since precise vertical position information is beyond the current capabilities of the global positioning system, approaches with LNAV/VNAV minimums make use of certified barometric VNAV (baro-VNAV) systems for vertical guidance and/or the wide area augmentation system (WAAS) to improve GPS accuracy for this purpose. (Note: WAAS makes use of a collection of ground stations that are used to detect and correct inaccuracies in the position information derived from the global positioning system. Using WAAS, the accuracy of vertical position information is increased to within 3 meters.) To make use of WAAS, however, the aircraft must be equipped with an IFR approved GPS receiver with WAAS signal reception that integrates WAAS error correction signals into its position determining processing. The WAAS enabled GPS receiver shown in *Figure 3-49* allows the pilot to load an RNAV approach and receive guidance along the lateral and vertical profile shown on the approach chart in *Figure 3-48*.

Figure 3-49. *WAAS data provides lateral and vertical guidance.*

It is very important to know what kind of equipment is installed in an aircraft, and what it is approved to do. It is also important to understand that the VNAV function of non-WAAS-capable or non-WAAS-equipped IFR approved GPS receivers does not make the aircraft capable of flying approaches to LNAV/VNAV minimums.

LPV

LPV can be thought of as "localizer performance with vertical guidance." Procedures with LPV minimums use GPS information to generate lateral guidance, and IFR-approved GPS/WAAS receivers to generate vertical guidance similar to an ILS glideslope. Several manufacturers now offer FMS/GPS RNAV units capable of flying approaches to LPV minimums.

GPS or RNAV (GPS) Approach Waypoints

Figure 3-50 shows a GPS approach loaded into an FMS/GPS RNAV. As previously noted, approaches must be selected from a specific approach menu in the FMS. The software then loads all of the waypoints associated with that procedure from the database into the flight route. It is not possible for you to enter or delete, separately or individually, waypoints associated with the approach procedure.

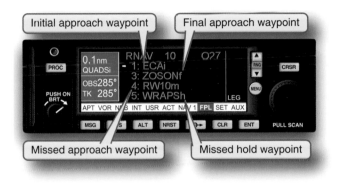

Figure 3-50. *Approach waypoints.*

Once loaded, a GPS or RNAV (GPS) approach is shown in the FMS display as a collection of waypoints with a title that identifies the approach. Four waypoints in every approach procedure have special designations: (1) initial approach waypoint; (2) final approach waypoint; (3) missed approach waypoint; and (4) missed holding waypoint.

Flying a GPS or RNAV (GPS) Approach

Most FMS require the pilot to choose whether to simply load, or load and activate, instrument approach procedures. When ATC tells you to expect a certain approach, select that approach from the menu and load it into the flight plan. Loading an approach adds its component waypoints to the end of the flight plan, but does not make them active. Once ATC clears you for the approach (or, alternatively, begins providing radar vectors to intercept the final approach course), you must remember to activate the approach to receive course guidance and auto-sequencing. You must be careful not to activate the approach until cleared to fly it, however, since activating the approach will cause the FMS to immediately give course guidance to the initial approach fix or closest fix outside the final fix, depending on the unit's programming. In the case of a vectors-to-final approach, activating the vector-to-final causes the FMS to draw a course line along the final approach course.

Once you have loaded and activated the GPS or RNAV (GPS) approach procedure, flying it is similar to flying between any other waypoints in a programmed flight route. However, you must be prepared for two important changes during the approach.

Terminal Mode

The first important change occurs when the aircraft reaches a point within 30 NM of the destination airport. At this point, regulations require that every GPS-based FMS/RNAV unit increase its sensitivity and integrity monitoring (receiver autonomous integrity monitoring, or RAIM, which continuously checks GPS signal reliability and alerts you if RAIM requirements are not met). If the system determines that RAIM requirements are met, the FMS/GPS RNAV unit automatically switches from en route sensitivity to terminal sensitivity within 30 NM of the destination airport. Terminal mode increases the sensitivity of the course deviation indicator (CDI) from 5 NM to 1 NM. The FMS/GPS RNAV displays an annunciation to let you know that it has switched from en route sensitivity to terminal mode.

Approach Mode

The second important change occurs 2 NM prior to reaching the final approach waypoint. At this point, the FMS/GPS

RNAV unit automatically switches to approach sensitivity. At this stage, the FMS/GPS RNAV further increases RAIM requirements, and increases the CDI sensitivity from 1 NM to 0.3 NM (i.e., a full-scale CDI deflection occurs if you are 0.3 NM or more from the desired course).

As long as the annunciation for approach mode is displayed, you may continue the approach. If, however, the computer fails to switch to approach mode, or the approach mode annunciation disappears, you must fly the published missed approach procedure. You are not authorized to descend further or to the MDA. Making changes to the FMS/GPS RNAV after reaching the 2 NM point could result in automatic cancellation of the approach mode.

Approach Not Active

If you arrive at the final approach waypoint and the approach mode is not active, you must fly the missed approach procedure. There should be no attempt to activate or reactivate the approach after reaching the final approach fix using any means—simply fly the missed approach procedure.

Vectored Approaches

As in conventional approaches, it is common for air traffic control to issue vectors to a GPS or RNAV (GPS) final approach course. Flying a vectored GPS or RNAV (GPS) approach is a simple matter of using the course intercept technique described in the previous section. The technique is illustrated again in *Figure 3-51*.

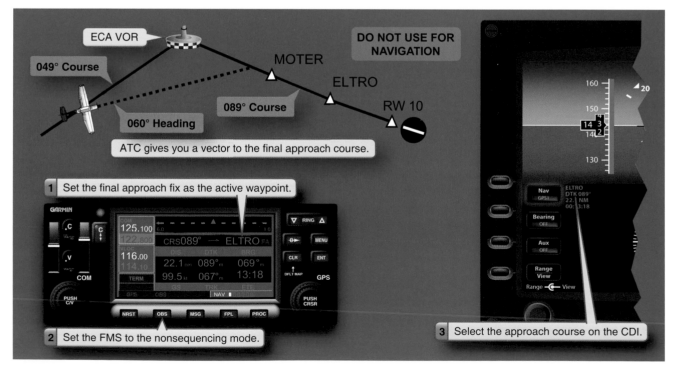

Figure 3-51. *A vectored RNAV approach.*

Many FMS/GPS RNAV units offer an automated solution to the problem of flying an approach in which the pilot receives vectors to the final approach course. Once ATC begins providing vectors to intercept the final approach course, you should activate the computer's "vectors-to-final" feature, which draws a course line along the final approach course. This feature helps you maintain situational awareness while being vectored because the assigned heading is clearly seen in relation to the final approach course. As already noted, you should monitor carefully to ensure that the FMS/GPS RNAV unit switches to approach mode within 2 NM of the final approach fix (FAF).

Figure 3-51 shows the vectors-to-final feature. The procedure required to use the vectors-to-final feature is illustrated in *Figure 3-52.*

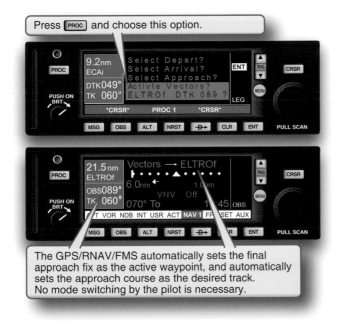

Figure 3-52. *The vectors-to-final feature.*

When set to use the vectors-to-final feature, many FMS/GPS RNAV units automatically set the FAF as the active waypoint; and set the final approach course as the desired track to the active waypoint.

Awareness: Briefing the Approach

As with any instrument approach, you should develop and consistently use a briefing technique to ensure that you think through all the steps necessary to set up the approach correctly. One technique uses the mnemonic ICE-ATM:

I   Identify primary navigation frequency

C   Course (inbound) set

E   Entry (direct, teardrop, parallel)

A   Altitudes for transition, initial, final, and missed approach segments

T   Timing/Distance(s)

M   Missed approach procedure

Another briefing technique uses the mnemonic FARS:

F   Frequencies set and identified

A   Altitudes for transition, initial, final, and missed approach   segments

R   Radial (inbound course) noted and set

S   Special notes (including missed approach procedure)

Common Error: Forgetting To Verify the Approach Mode

The most common error made during a GPS/RNAV approach is to forget to ensure that the approach mode has indeed engaged prior to beginning a descent to minimums. Routinely checking for the approach indication 2 NM before the final approach waypoint not only prevents this type of error, but also gives you a minute or so to remedy some situations in which the approach mode has not engaged.

Common Error: Using the Wrong Approach Minimums

Listing several different approach minimums on a single instrument approach chart introduces the possibility of another simple type of error: using the wrong approach minimums. One way to avoid mix-ups is to verbalize the equipment being used and type of procedure being flown, and then search for the approach minimums with these details in mind. You must be absolutely certain of the certification, approval, and installed options of the advanced avionics equipment prior to flight planning.

Common Error: Forgetting To Reengage Sequencing Mode Prior to Final Approach Waypoint

A common mistake made by pilots when they are learning to fly vectored approaches without a vectors-to-final feature is forgetting to set the FMS/GPS RNAV back to sequencing mode once established on the approach course. This error prevents the FMS/ GPS RNAV unit from switching to the approach mode 2 NM prior to the FAF. If you pass the final approach fix and the computer is still in nonsequencing mode, the approach mode will be disabled and you must fly the missed approach, report the missed approach and request another approach.

Essential Skills

1. Load and activate a vectored GPS or RNAV (GPS) approach.

2. Select a vectored initial approach segment.

3. Determine the correct approach minimums and identify all pertinent mode transitions.

4. Determine the published missed approach point (MAP), courses, altitudes, and waypoints to fly.

5. Determine how missed approach guidance is selected.

Course Reversals

Figure 3-53 shows three common course reversals: (1) 45-degree procedure turn, (2) holding pattern, and (3) teardrop procedure.

Course reversals are handled in the same way as holding procedures, by using the FMS/GPS's nonsequencing mode. As you arrive at the initial approach waypoint, the unit's nonsequencing mode should be engaged to prevent it from immediately sequencing to the next waypoint in the approach. After completing the course reversal, be sure to re-engage the system's sequencing mode to continue the approach.

The navigation unit in *Figure 3-54* requires that you manually switch between the sequencing and nonsequencing modes.

Preprogrammed Course Reversals

Some FMS/GPS units insert preprogrammed course reversals into published instrument approach procedures. The purpose of a preprogrammed course reversal is to relieve you from the mode switching and course selection tasks associated with course reversals.

The FMS/GPS unit in *Figure 3-55* includes a preprogrammed course reversal. This unit automatically sets the outbound course for the outbound portion of the course reversal. Once the turn inbound has been made, the unit automatically sets the inbound course back to the final approach waypoint.

This FMS/GPS unit does not switch between sequencing and nonsequencing modes for a 45-degree course reversal (although it does for a holding-type course reversal). Whether it is done manually, automatically, or not at all, you must be sure that the system is engaged in sequencing mode before reaching the final approach waypoint after the course reversal is completed. The FMS/GPS will switch to the approach mode only if the system is engaged in the sequencing mode.

Common Error: Mismanaging the Sequencing/Nonsequencing Modes During a Course Reversal

Neglecting to switch the FMS/GPS from the nonsequencing mode prior to reaching the initial approach waypoint and neglecting to switch the system back to the sequencing mode prior to passing the final approach waypoint are common errors made during course reversals.

Essential Skills

1. Select a type of course reversal procedure.

2. Determine the correct sequence of mode control actions to be accomplished by the pilot.

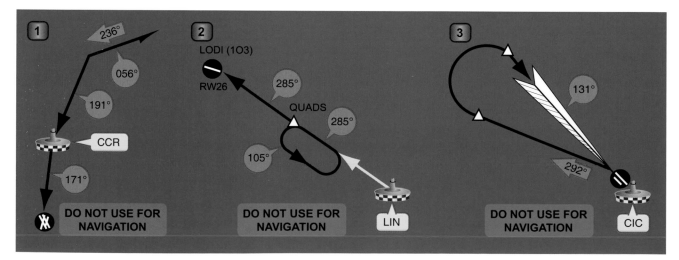

Figure 3-53. *Three types of course reversals.*

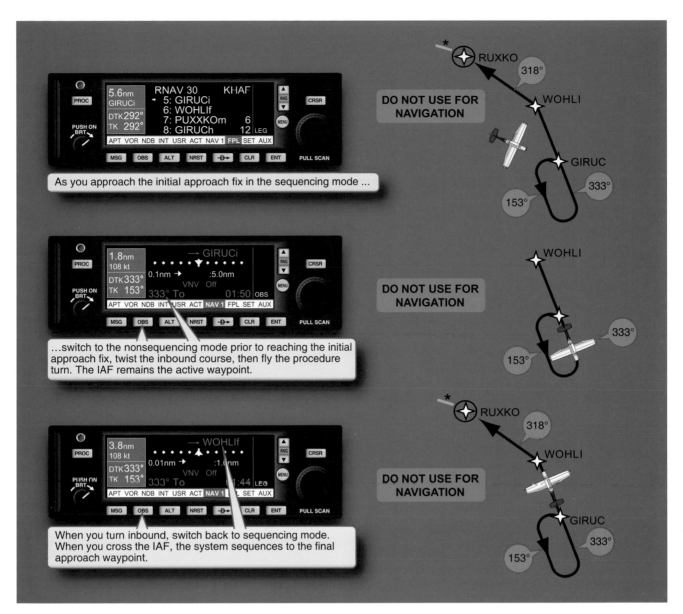

Figure 3-54. *Using the nonsequencing mode to accomplish a course reversal.*

As you approach the initial approach fix in the sequencing mode ...

...switch to the nonsequencing mode prior to reaching the initial approach fix, twist the inbound course, then fly the procedure turn. The IAF remains the active waypoint.

When you turn inbound, switch back to sequencing mode. When you cross the IAF, the system sequences to the final approach waypoint.

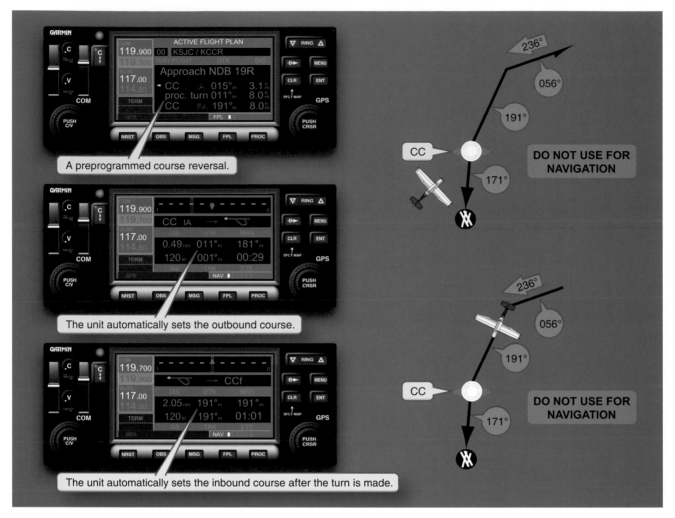

Figure 3-55. *Using the nonsequencing mode to accomplish a course reversal.*

Missed Approaches

The FMS/GPS unit's nonsequencing mode provides an easy way to fly missed approach procedures, such as the one illustrated in *Figure 3-56*.

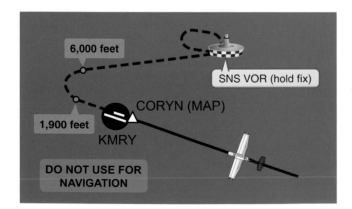

Figure 3-56. *A missed approach procedure.*

The missed approach procedure shown in *Figure 3-56* requires you to climb to 1,900 feet, turn right and climb to 6,000 feet, then proceed direct to the SNS VOR.

The FMS/GPS helps you navigate between waypoints, which are geographically fixed locations. But where will the aircraft reach 1,900 feet on the missed approach procedure at Monterey? This depends on what aircraft you are flying and the chosen rate of climb. A single-engine airplane might be four miles away by the time it reaches 1,900 feet. A small jet might reach 1,900 feet by the end of the runway. The problem is that, given the way the FMS/GPS system uses waypoints, there is no one way to represent the climbs and turns required on a missed approach procedure.

To address this issue, all FMS/GPS RNAV units automatically suspend waypoint sequencing when you reach the missed approach point. The unit waits until you acknowledge the passing of the MAP before it continues the sequencing. When the aircraft has gained the published altitudes and complied with the initial MAP procedures, you can safely proceed to the missed approach holding waypoint, being mindful of any altitude requirements. A waypoint for the missed approach holding point is included as part of the missed approach procedure. In the example above, you can make the missed approach holding waypoint the active waypoint, and re-engage the sequencing mode upon reaching 6,000 feet. You now have sequencing mode guidance to the missed approach holding waypoint. The procedure for one FMS/GPS is illustrated in *Figure 3-57*.

Since the hold at SNS is part of the published missed approach procedure, it can be carried out using the same technique used to perform a holding pattern. Some FMS/GPS units will automatically switch to the nonsequencing mode when you reach the hold fix. Other units may advise you to switch manually to the nonsequencing mode.

Recognizing the Missed Approach Point

With any type of navigation equipment, it is important to be able to determine when you have reached the missed approach point. The missed approach point indications given by FMS/GPS units are sometimes subtle. Consider the two navigation displays shown in *Figure 3-58*. The display in the

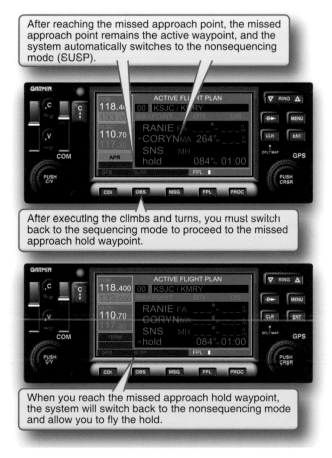

Figure 3-57. *Flying a missed approach procedure.*

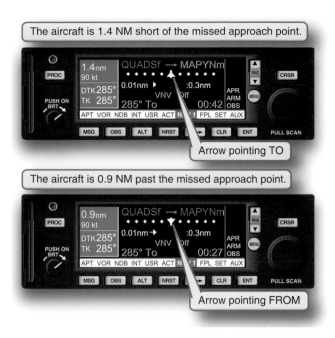

Figure 3-58. *Recognizing the missed approach waypoint.*

top graphic of *Figure 3-58* shows the aircraft approaching the missed approach point, 1.4 NM away.

Now consider the display in the bottom graphic of *Figure 3-58*. The distance from the missed approach point might suggest that the aircraft is now even closer to the missed approach point. However, the TO/FROM flag on the course deviation indicator shows that the aircraft has in fact passed the missed approach point. It is tempting to monitor the distance from the missed approach point as it decreases to 0.0 NM. The problem is that, depending on how accurately the pilot flies, the distance may never reach 0.0 NM. Rather, it may simply begin to increase once you have passed abeam the missed approach point. It is, thus, important to check not only the distance from the missed approach point, but also the TO/FROM flag or arrow. In the rush of a missed approach, this small clue (arrow direction change) can be difficult to read and very easy to misinterpret.

Complying With ATC-Issued Missed Approach Instructions

ATC sometimes issues missed approach instructions that are different from those published on the approach chart. In this case, use the techniques described earlier to insert new waypoints into the route, and/or to intercept and track courses to those waypoints.

Setting Up Next Procedure in Hold

Once in the missed approach holding pattern, the next task is deciding where to go next and programming the new flight plan into the FMS/GPS unit. In this high workload situation, it is especially important to be very proficient with the menus, functions, and "switchology" of a particular unit. If the aircraft is equipped with an autopilot, it is also essential to have a thorough understanding of how the autopilot interacts and interfaces with the FMS/GPS navigation equipment.

Common Error: Noncompliance With Initial Missed Approach Instructions

The immense capability of the FMS/GPS may tempt you to follow its directions rather than fly a missed approach procedure exactly as published on the instrument approach procedure chart. Always fly procedures as published, especially with respect to the initial climb and turn instructions. GPS as a line-of-sight navigation aid can display courses and distances to a ground-based navaid even though the navaid is on the other side of a mountain range and itself cannot be received, because GPS signals are spaced based.

Essential Skills

1. Acknowledge a missed approach procedure.

2. Set the FMS/GPS for a return to the same approach to fly it again.

3. Select a different approach while holding at a missed approach holding waypoint.

4. Program an ATC specified hold (user waypoint) point for selection after the published MAP/hold procedure.

Ground-Based Radio Navigation

Configuring FMS To Receive Ground-Based Radio Navigation Signals

Most advanced avionics systems include receivers for conventional radio navigation signals from VOR, localizer, and glideslope transmitters. To display these signals on the navigation display indicator(s), you need two fundamental skills.

Tuning and Identifying Radio Navigation Facilities

The first fundamental skill in ground-based radio navigation is tuning and identifying a ground-based radio navigation facility. *Figure 3-59* illustrates how a VOR station can be tuned using two different systems.

Some systems automatically attempt to identify ground-based radio navigation facilities that are selected by the pilot. Note the identifier that appears beside the selected frequency in the upper left corner of the PFD in *Figure 3-59* (116.00 = ECA).

Displaying Radio Navigation Signals on the Navigation Indicator

The second fundamental skill is displaying indications from a ground-based radio navigation facility on the navigation display indicator in the aircraft. In addition to setting the navigation indicator to display indications from different navigation sources, you must also know where to look to double-check which indications are currently being displayed. It is crucial to remain constantly aware of the navigation source for each indicator. Many systems use color coding to make a visual distinction between different RNAV navigation sources (GPS, INS, etc.) and ground-based radio navigation sources.

Awareness: Using All Available Navigation Resources

Looking at the two systems shown in *Figure 3-59,* you can see that two VOR frequencies appear in the active windows at all times, regardless of whether VOR or GPS is being used as the primary navigation source. To maximize situational awareness and make best use of this resource, it is a good practice to keep them tuned to VOR stations along your route of flight. If you have two navigation indicators, you can have one indicator set to show GPS course indications, with the other to show VOR indications. Used in this way, VOR and GPS can serve as backups for each another.

Flying a Precision Approach Using Ground-based Navigation Facilities

Flying a precision approach requires tuning the required frequencies, configuring the navigation indicator to display localizer course indications, and flying the approach. For aircraft equipped with multiple navigation radios, the localizer frequency can go into one receiver, while a second navigational facility used as a cross-radial can be set in the other receiver. As you come within range of the localizer and glideslope, the course deviation and glideslope indicators will show position with respect to the localizer and glideslope.

Flying a Nonprecision Approach Using Ground-Based Navigation Facilities

Nonprecision approaches such as VOR, localizer, and LDA approaches are flown using the same procedures used to fly a precision approach. If the aircraft is equipped with an autopilot, be sure to develop a thorough understanding of how the autopilot works with the FMS. While these systems automate some tasks, others (e.g., flying the procedure turn course reversal) maybe left to the pilot.

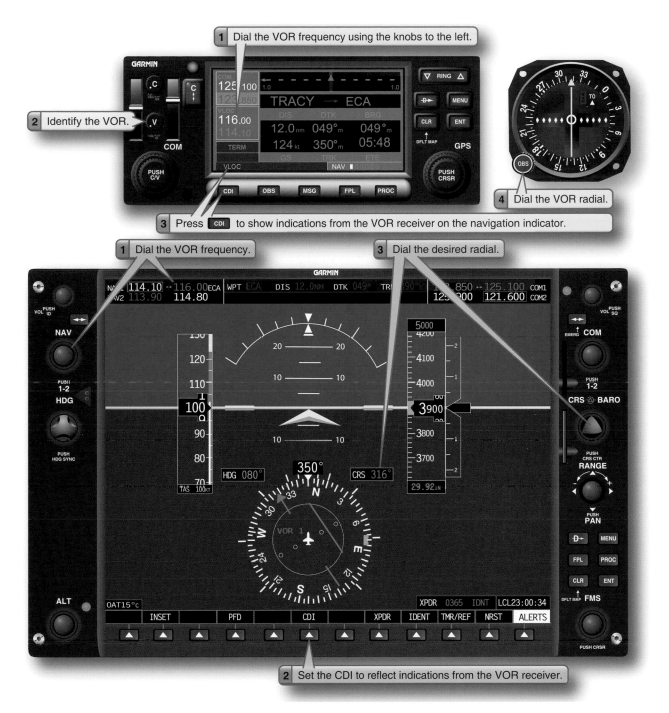

Figure 3-59. *Tuning navigation radio frequencies.*

Maintaining Proficiency: Practicing All Navigation Skills

Advanced avionics systems offer you several ways to navigate. Numerous studies have demonstrated the potential for deterioration of navigational skills that are not regularly practiced. It is important to get regular practice using ground-based navigation facilities as well as RNAV sources. One way to maintain proficiency is to consistently use ground-based navigational facilities as a backup to RNAV systems.

Essential Skills

1. Select any type of ground-based radio navigation approach.

2. Correctly tune and set up the conventional navigation receiver for that procedure.

3. Correctly monitor the navaid for properly identification and validity.

4. Correctly select and be able to use the desired navigation source for the autopilot.

Chapter Summary

Navigation has been freed from the constraints of channeling all flight traffic along one path. The area navigation capabilities found in advanced avionics receiving signals from other than conventional line-of-sight ground-based aviation navaids and the compact size and reliability of microchips now allow efficient, accurate air travel. Integrated databases facilitated by large reliable memory modules help you to select routes, approaches, and avoid special use airspace.

With this freedom of movement, you must expend more time learning the system and how to do the preflight entries or programming. In addition to current charts, you must now verify the currency of the advanced avionics databases. The aircraft owner must also allocate the funding to maintain the currency of the databases.

You now have access to a tremendous amount of data. The methods of data selection and display must be learned and then decisions made about which display formats to use at which times. VOR/DMEs are simple receivers to tune and use. To use current flight management systems and area navigation units, you may need to study books that are larger than the actual units themselves. You must know the quality of maintenance for advanced avionics units and the qualifications of the systems to determine appropriate uses of the equipment.

Since advanced avionics have different displays, navigation sources, functions, and features, the pilot must always be aware of the mode selected, the data source(s), and the function selected. Pilot lack of attention to navigation can have dire consequences, including notification of the next of kin.

Chapter 4
Automated Flight Control

Introduction

This chapter introduces automated flight control in the advanced avionics cockpit. You will learn to use an autopilot system that can significantly reduce workload during critical phases of flight. The two-axis autopilot system installed in most general aviation aircraft controls the pitch and roll of the aircraft. The autopilot can operate independently, controlling heading and altitude, or it can be coupled to a navigation system and fly a programmed course or an approach with glideslope. In addition to learning how to use the autopilot, you must also learn when to use it and when not to use it.

Press the APR button to engage the approach func[...]

The autopilot should maintain a greater accuracy of track in the approach mode.

autopilot control panel can be used
[...]ed altitudes and vertical speeds.

The CDI is coupled to the VOR receiver, so the NAV function
steers the aircraft along the desired track to the selected radial.

You will learn how the autopilot and the flight management system (FMS)/area navigation (RNAV) unit combine to create a fairly automated form of flight that places you in a managerial role. While the autopilot relieves you from manually manipulating the flight controls, you must maintain vigilance over the system to ensure that it performs the intended functions and the aircraft remains within acceptable parameters of altitudes, airspeeds, and airspace limits.

Autopilot Concepts

An autopilot can be capable of many very time intensive tasks, helping the pilot focus on the overall status of the aircraft and flight. Good use of an autopilot helps automate the process of guiding and controlling the aircraft. Autopilots can automate tasks, such as maintaining an altitude, climbing or descending to an assigned altitude, turning to and maintaining an assigned heading, intercepting a course, guiding the aircraft between waypoints that make up a route programmed into an FMS, and flying a precision or nonprecision approach. You must accurately determine the installed options, type of installation, and basic and optional functions available in your specific aircraft.

Many advanced avionics installations really include two different, but integrated, systems. One is the autopilot system, which is the set of servo actuators that actually do the control movement and the control circuits to make the servo actuators move the correct amount for the selected task. The second is the flight director (FD) component. The FD is the brain of the autopilot system. Most autopilots can fly straight and level. When there are additional tasks of finding a selected course (intercepting), changing altitudes, and tracking navigation sources with cross winds, higher level calculations are required.

The FD is designed with the computational power to accomplish these tasks and usually displays the indications to the pilot for guidance as well. Most flight directors accept data input from the air data computer (ADC), Attitude Heading Reference System (AHRS), navigation sources, the pilot's control panel, and the autopilot servo feedback, to name some examples. The downside is that you must program the FD to display what you are to do. If you do not preprogram the FD in time, or correctly, FD guidance may be inaccurate.

The programming of the FD increases the workload for the pilot. If that increased workload is offset by allowing the autopilot to control the aircraft, then the overall workload is decreased. However, if you elect to use the FD display, but manually fly the aircraft, then your workload is greatly increased.

In every instance, you must be absolutely sure what modes the FD/autopilot is in and include that indicator or annunciator in the crosscheck. You must know what that particular mode in that specific FD/autopilot system is programmed to accomplish, and what actions will cancel those modes. Due to numerous available options, two otherwise identical aircraft can have very different avionics and autopilot functional capabilities.

How To Use an Autopilot Function

The following steps are required to use an autopilot function:

1. Specify desired track as defined by heading, course, series of waypoints, altitude, airspeed, and/or vertical speed.

2. Engage the desired autopilot function(s) and verify that, in fact, the selected modes are engaged by monitoring the annunciator panel.

3. Verify that desired track is being followed by the aircraft.

4. Verify that the correct navigation source is selected to guide the autopilot's track.

5. Be ready to fly the aircraft manually to ensure proper course/clearance tracking in case of autopilot failure or misprogramming.

6. Allow the FD/autopilot to accomplish the modes selected and programmed without interference, or disengage the unit. Do not attempt to "help" the autopilot perform a task. In some instances this has caused the autopilot to falsely sense adverse conditions and trim to the limit to accomplish its tasking. In more than a few events, this has resulted in a total loss of control and a crash.

Specification of Track and Altitude

A track is a specific goal, such as a heading or course. A goal can also be a level altitude, a selected airspeed, or a selected vertical speed to be achieved with the power at some setting. Every autopilot uses knobs, buttons, dials, or other controls that allow the pilot to specify goals. *Figure 4-1* shows an autopilot combined with conventional navigation instruments. Most autopilots have indicators for the amount of servo travel or trim being used. These can be early indicators of adverse conditions, such as icing or power loss. Rarely will a trim indicator ever indicate full travel in normal operation. Consistently full or nearly full travel of the trim servos may be a sign of a trim servo failure, a shift in weight resulting in a balance problem, or airfoil problems such as icing or inadvertent control activation.

A heading knob on the direction indicator is used to specify target headings.

The OBS knob on the CDI is used to specify target courses.

Knobs on the autopilot control panel can be used to enter assigned altitudes and vertical speeds.

Figure 4-1. *A simple autopilot.*

Primary flight displays (PFDs) often integrate all controls that allow modes to be entered for the autopilot. The PFD shown in *Figure 4-2* offers knobs that allow you to enter modes without turning attention away from the primary flight instruments. Modes entered using the controls on a PFD are transferred to the autopilot.

Engagement of Autopilot Function

Every autopilot offers a collection of buttons that allow you to choose and engage autopilot modes and functions. Buttons used to engage autopilot modes appear along the bottom of the autopilot shown in *Figure 4-1*. The system shown in *Figure 4-3* does not use a separate device for autopilot controls; it integrates the autopilot function buttons into another cockpit display.

Verification of Autopilot Function Engagement

It is very important to verify that an autopilot mode has engaged, and the aircraft is tracking the intended flight

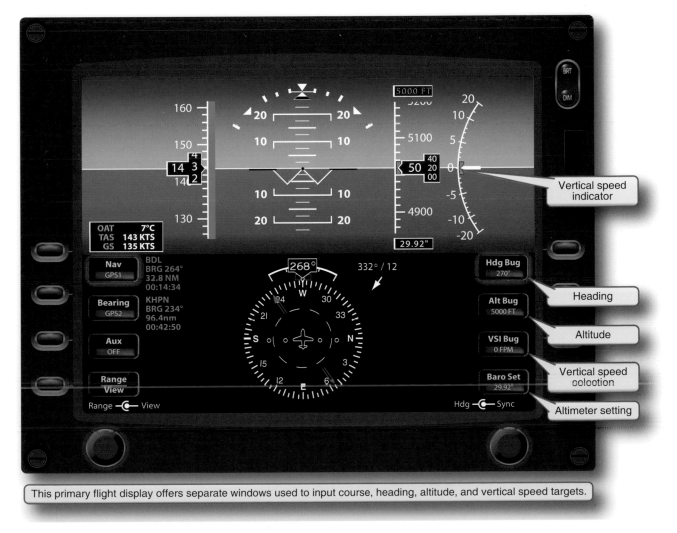

This primary flight display offers separate windows used to input course, heading, altitude, and vertical speed targets.

Figure 4-2. *Entering goals on a primary flight display.*

Figure 4-3. *An integrated avionics system with an autopilot.*

profile. Every autopilot displays which autopilot modes are currently engaged, and most indicate an armed mode that activates when certain parameters are met, such as localizer interception. The autopilot shown in *Figure 4-1* displays the active modes on the front of the unit, just above the controls. The integrated autopilot shown in *Figure 4-4* displays the currently engaged autopilot mode along the top of the PFD.

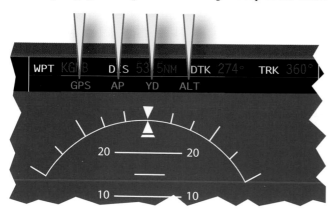

Figure 4-4. *Engaged autopilot modes shown at the top of a PFD.*

How Autopilot Functions Work

Once an autopilot mode has been engaged, the autopilot:

1. Determines which control movements are required to follow the flight profile entered by the pilot, and

2. Moves the controls to affect tracking of the flight profile.

Determination of Control Movements Required To Achieve Goals

Suppose you wish to use the autopilot/FD to turn to an assigned heading of 270°. The heading knob is used to select the new heading. Before any control movements are made, the autopilot/FD must first determine which control movements are necessary (e.g., left or right turn). To do so, the FD/autopilot must first determine the aircraft's current heading and bank angle, determine amount and direction of the turn, and then choose an appropriate bank angle, usually up to 30° or less. To make these determinations, the FD gathers and processes information from the aircraft's ADC (airspeed and altitude), magnetic heading reference instrument, and navigation systems.

Carrying Out Control Movements

Once the FD/autopilot has determined which control movements are necessary to achieve the flight change, the autopilot has the task of carrying out those control movements. Every autopilot system features a collection of electromechanical devices, called servos, that actuate the aircraft control surfaces. These servos translate electrical commands into motion, the "muscle" that actually moves the control surfaces.

Flight Director

Flight Director Functions

An FD is an extremely useful aid that displays cues to guide pilot or autopilot control inputs along a selected and computed flightpath. *[Figure 4-5]* The flight director usually receives input from an ADC and a flight data computer. The ADC supplies altitude, airspeed and temperature data, heading data from magnetic sources such as flux valves, heading selected on the HSI (or PFD/multi-function display (MFD)/ electronic horizontal situation indicator (EHSI)), navigation data from FMS, very high frequency omnidirectional range (VOR)/distance measuring equipment (DME), and RNAV sources. The flight data computer integrates all of the data such as speed, position, closure, drift, track, desired course, and altitude into a command signal.

The command signal is displayed on the attitude indicator in the form of command bars, which show the pitch and roll

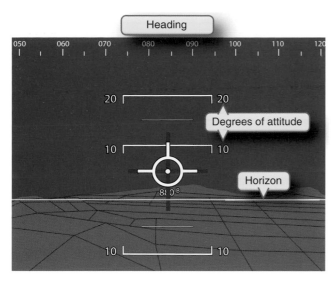

Figure 4-5. *A flight director.*

inputs necessary to achieve the selected targets. To use the flight director command bars, which are usually shaped as inverted chevrons, or V-shaped symbols, the pilot simply flies to the bars. Some older models use crossed bars, leading the pilot to the selected point. In both types, you simply keep the aircraft symbol on the attitude indicator aligned with the command bars, or allow the autopilot to make the actual control movements to fly the selected track and altitude.

Using the Flight Director (FD)
Flight Director Without Autopilot

The FD and autopilot systems are designed to work together, but it is possible to use the flight director without engaging the autopilot, or the autopilot without the FD, depending on the installation. Without autopilot engagement, the FD presents all processed information to the pilot in the form of command bar cues, but you must manually fly the airplane to follow these cues to fly the selected flightpath. In effect, you "tell" the FD what needs to happen and the FD command bars "tell" you what to do. This adds to your workload, since you must program the FD for each procedure or maneuver to be accomplished, while actually flying the aircraft. In many cases, you will have a decreased workload if you simply disable the FD and fly using only the flight instruments.

Flight Director With Autopilot

When the aircraft includes both a flight director and an autopilot, you may elect to use flight director cues without engaging the autopilot. It may or may not be possible to use the autopilot without also engaging the flight director. You need to be familiar with the system installed. When you engage the autopilot, it simply follows the cues generated by the flight director to control the airplane along the selected lateral and vertical paths.

Common Error: Blindly Following Flight Director Cues

The convenience of flight director cues can invite fixation or overreliance on the part of the pilot. As with all automated systems, you must remain aware of the overall situation. Never assume that flight director cues are following a route or course that is free from error. Rather, be sure to include navigation instruments and sources in your scan. Remember, the equipment will usually perform exactly as programmed. Always compare the displays to ensure that all indications agree. If in doubt, fly the aircraft to remain on cleared track and altitude, and reduce automation to as minimal as possible during the problem processing period. The first priority for a pilot always is to fly the aircraft.

Common Error: Confusion About Autopilot Engagement

Pilots sometimes become confused about whether or not flight director cues are being automatically carried out by the autopilot, or left to be followed manually by the pilot. Verification of the autopilot mode and engagement status of the autopilot is a necessary technique for maintaining awareness of who is flying the aircraft.

Follow Route

The FD/autopilot's navigation function can be used to guide the aircraft along the course selected on the navigation indicator. Since the navigation display in most advanced avionics cockpits can present indications from a variety of navigation systems, you can use the autopilot's navigation function to follow a route programmed into the FMS using VOR, global positioning system (GPS), inertial navigation system (INS), or other navigation data sources.

Following a Route Programmed in the FMS

Figure 4-6 demonstrates how to use the navigation function to follow a route programmed into the FMS. With the navigation function engaged, the FD/auto-pilot steers the aircraft along the desired course to the active waypoint. Deviations from the desired course to the new active waypoint are displayed on the navigation indicator. When the aircraft reaches the active waypoint, the FMS computer automatically sequences to the next waypoint in the route, unless waypoint sequencing is suspended.

It is important to note that the normal navigation function provides only lateral guidance. It does not attempt to control the vertical path of the aircraft at any time. You must always ensure the correct altitude or vertical speed is maintained.

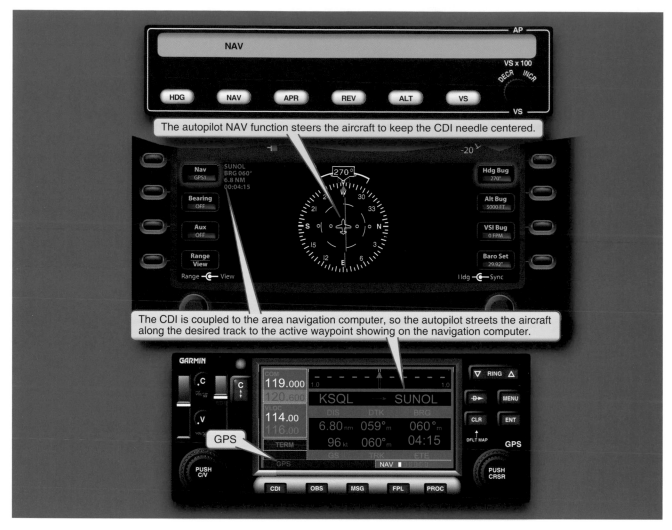

Figure 4-6. *Using the navigation function to follow the programmed flight route.*

When combined, use of the FMS and the FD/autopilot's navigation function result in an automated form of flight that was formerly limited to very complex and expensive aircraft. This same level of avionics can now be found in single-engine training airplanes. While it is easy to be complacent and let down your guard, you must continuously monitor and stay aware of automated systems status and function and the track of the aircraft in relation to the flight plan and air traffic control (ATC) clearance.

GPS Steering (GPSS) Function

Many autopilots offer a global positioning system steering (GPSS) function. GPSS does all of the same actions as the navigation function, but achieves a higher degree of precision by accepting inputs directly from the GPS receiver. Consequently, the GPSS function follows the desired track to the active waypoint more aggressively, permitting only small excursions from the desired course. On some installations, pressing the autopilot NAV button twice engages the GPSS function.

Following a VOR Radial

The FD/autopilot's navigation function can also be used to directly track VOR radials. The navigation display must be configured to show indications from one of the aircraft's VOR receivers. Once you have tuned and identified a VOR station and selected the desired radial, you can select the navigation mode to track the selected radial. *Figure 4-7* demonstrates how to use the navigation mode to follow a VOR radial.

When the navigation mode is used to follow a route defined by VOR radials, you must still tune and identify each new VOR facility manually and select the appropriate radials along the way. The autopilot's navigation function cannot automatically manipulate the VOR receiver. However, some highly automated FMS units tune and identify VORs along a defined route, such as Victor or Jet routes. You should check the FMS documentation and installed options.

Figure 4-7. *Using the navigation mode to follow a VOR radial.*

Depending on the FMS, the highly automated flight that results when the navigation mode is used to follow a published route from the database uses a different skill set from using the navigation mode to track discreetly tuned VOR radials. Learning how to select preprogrammed routes from the database of airways can be challenging. Programming or tuning discreet VORs en route in turbulent conditions presents different challenges. Either skill set can result in a greater sharing of duties between pilot and technology and an increase in safety.

Fly Heading

The heading mode is used to steer the aircraft automatically along a pilot selected heading. Using the FD/autopilot to fly a heading is a simple matter of selecting the assigned heading and then engaging the heading function or, more commonly, accomplished by first engaging the heading mode and gently turning the heading selection knob to the new heading. Gently turning the knob with the mode already engaged allows you to make a smooth change from level to turning flight. Many

autopilots make an abrupt bank if engaged when there is a big change to be made in heading or track. The heading function is illustrated in *Figure 4-8*.

You should note that, when using the heading mode, the FD/autopilot ignores the pilot-programmed route in the FMS or any VOR radials you set. When in heading mode, the FD/autopilot will fly the selected heading until fuel starvation.

Maintain Altitude

The autopilot's altitude mode maintains an assigned barometric altitude. When the altitude mode is engaged, the autopilot seeks to maintain the same barometric pressure (altitude) that the aircraft was flying at the time that the altitude mode was engaged. *Figure 4-9* shows how to engage the altitude mode for one manufacturer's autopilot.

In addition to determining and carrying out the pitch commands necessary to maintain the flight's assigned altitude, most autopilots are also able to trim the aircraft.

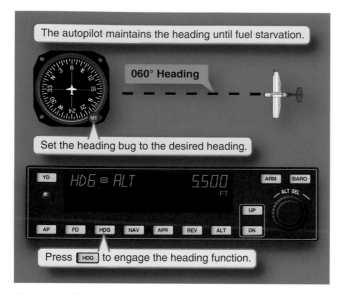

Figure 4-8. *Flying an assigned heading using the heading mode.*

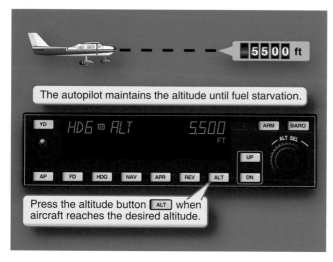

Figure 4-9. *Maintaining an altitude using the altitude hold mode.*

An autotrim system is capable of automatically making any needed adjustments to the pitch trim to maintain the aircraft at the desired altitude and in a properly trimmed condition. Pitch control pressure applied with the altitude hold mode engaged will cause the autopilot to trim against you.

Climbs and Descents

Vertical Speed

The autopilot's vertical speed mode allows you to perform constant-rate climbs and descents. *Figure 4-10* illustrates the use of the vertical speed mode for one autopilot that is integrated with a PFD.

When you engage the vertical speed mode, the FD/autopilot will attempt to maintain the specified vertical speed until you choose a different setting in autopilot, the aircraft reaches an assigned altitude set into the assigned altitude selector/alerter,

or the autopilot is disconnected. If an altitude selector is not installed or functioning, the pilot has the task of leveling off at the assigned altitude, which requires monitoring progress and manually engaging the autopilot's altitude hold function once the aircraft reaches the desired altitude. You must be very careful to specify an appropriate vertical speed, as the aircraft will fly itself into a stall if you command the autopilot to climb at a rate greater than the aircraft's powerplant(s) is/are capable of supporting. You also need to monitor descent airspeeds diligently to ensure compliance with V_{NE}/V_{MO} and V_A or turbulence penetration speeds if there is doubt about smooth air conditions. As discussed in the previous chapter, you should be cognizant of the powerplant temperatures reciprocating powered aircraft and bleed air requirements for turbine-powered aircraft.

Vertical Speed With Altitude Capture

Some FD/autopilots have an altitude select/capture feature. The altitude select/capture feature is illustrated in *Figure 4-11*. The altitude select/capture feature combines use of the activated vertical speed mode and an armed altitude hold mode. To use this feature, the vertical speed function is initially engaged. The altitude hold mode usually arms automatically when a different altitude is selected for capture and vertical speed is activated. With an altitude select/capture option or feature, the altitude hold mode disengages the vertical speed mode upon capture of the selected altitude once the vertical speed function completes the necessary climb or descent. Once the aircraft reaches the assigned altitude, the vertical speed function automatically disengages, and the altitude mode changes from armed to engaged. The change from vertical speed mode to altitude hold mode is the capture mode, or transition mode. Any changes made by the pilot during this short phase usually result in a cancellation of the capture action, allowing the aircraft to continue the climb or descent past the selected altitude. Again, be familiar with the aircraft's equipment. Let the system complete programmed tasks, and understand what it will do if interrupted.

Many FD/autopilot altitude selectors include an altitude alert feature, an auditory alert that sounds or chimes as the aircraft approaches or departs the selected altitude.

Catching Errors: Armed Modes Help Prevent Forgotten Mode Changes

You have already seen how remembering to make a needed mode change in the future can be an error-prone process. Not canceling the armed function allows the altitude select mode to relieve the pilot from needing to remember to engage the function manually once the aircraft has reached the selected altitude. Do not interrupt the altitude armed or capture mode, unless prepared to manually control the process.

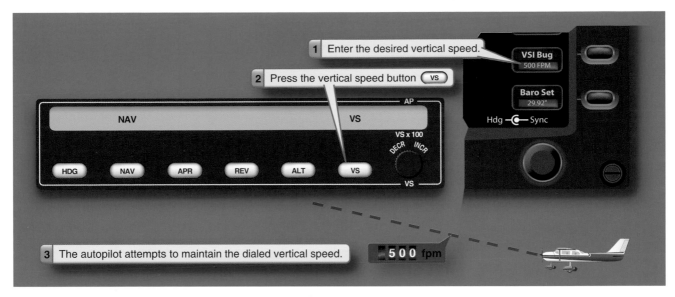

Figure 4-10. *Performing a constant-rate climb or descent using the vertical speed function.*

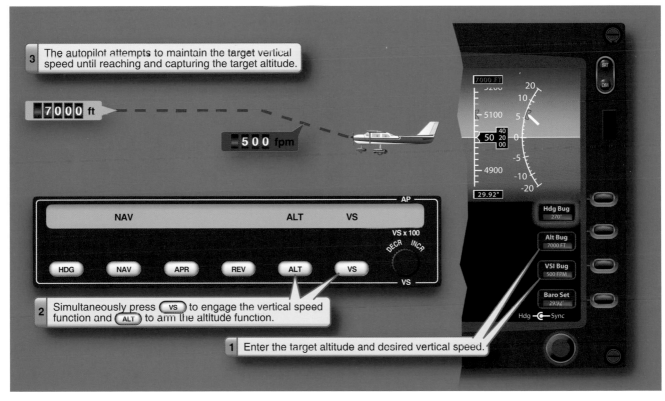

Figure 4-11. *Climbs and descents to capture using the altitude select/capture feature*

The indications on the autopilot in *Figure 4-11* do not distinguish between functions that are armed or engaged.

The more sophisticated annunciator shown in *Figure 4-12* uses color coding to distinguish between armed and engaged autopilot functions.

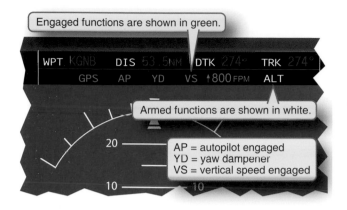

Figure 4-12. *A mode annunciator showing armed and engaged autopilot modes.*

Common Error: Failure To Arm the Altitude Mode

The most common error made by pilots during climbs and descents is failure to arm the altitude mode to capture the assigned altitude. In many instances, this happens when the crew does not correctly adjust the altitude selector or alerter. Sometimes, this malfunction occurs when the altitude is adjusted at the same time the system is attempting to go into the capture mode. This situation typically results in the aircraft climbing or descending beyond the assigned altitude, which may result in an altitude deviation. Altitude deviations are among the most common mishaps reported by pilots to NASA's Aviation Safety and Reporting System (ASRS). In any event, always monitor the actions of the FD/autopilot system and be prepared to fly the aircraft manually.

Awareness: Altitude Alerting Systems

Altitude alerting systems were mandated for commercial jet transports in the early 1970s in response to a growing number of altitude deviations in airline operations. Although they helped reduce the total number of altitude deviations, altitude alerting systems also made possible a new kind of error. Altitude deviation reports submitted to the Aviation Safety and Reporting System (ASRS) indicate that pilots sometimes rely too much on the altitude alerting system, using it as a substitute for maintaining altitude awareness. Instead of monitoring altitude, pilots sometimes simply listen for the alert. This phenomenon is one instance of what human factors experts call primary-secondary task inversion—when an alert or alarm designed as a secondary backup becomes the primary source of information. In the case of the altitude

alerting system, when the alerting system is missed, or you are distracted, nothing is left to prevent an altitude deviation. You must remember that the altitude alerting system is designed as a backup, and be careful not to let the alerting system become the primary means of monitoring altitude. Most airline operators have a standard operating procedure that requires pilots to call out approaching target altitudes before the altitude alerting system gives the alert. Common errors occur when setting 10,000 feet versus 11,000 feet. Too many ones and zeros can confuse a fatigued, busy pilot, resulting in setting an incorrect altitude.

Awareness: Automatic Mode Changes

Distinguishing between "armed" and "engaged" adds complexity to the process of maintaining mode awareness. In addition to autopilot functions that are engaged by the pilot, some autopilot functions engage and disengage automatically. Automatic mode changes add to the challenge of keeping track of which autopilot functions are currently engaged and which functions are set to become engaged. You can minimize confusion by always verifying the status annunciations on the FMs, PFD/MFD, and the autopilot mode annunciator after any change of heading, altitude, or vertical speed. The verification process forces you to carefully consider the configuration of the FMS and FD/autopilot. Determine if engaging the autopilot cancels certain FD modes. Some units interact, and when the autopilot is engaged, some FD modes are automatically canceled, notably altitude hold or selection.

Learning: The Importance of Understanding

One way to learn the steps required to use an autopilot is simply to memorize them. This approach focuses solely on the button and control manipulations required to perform each procedure. Although this approach to learning may appear to be the quickest, studies have shown that pilots who take the time to develop a deeper understanding of how a system works give themselves three important advantages. These pilots are better able to:

1. Work through situations that differ from the ones they learned and practiced during training,

2. Transition from one manufacturer's system to another, and

3. Recall procedures after not having practiced them for some time.

Investing time to understand FD/autopilot functions pays off. For example, in many systems, once the aircraft reaches the selected altitude and levels off as indicated by the altitude mode annunciator, the pilot can select the next altitude in the window. Then, upon receiving the clearance to climb or descend, the pilot must select only the vertical mode. In many systems, the vertical speed mode is indicated and the altitude mode is indicated as "armed" and ready to capture the selected altitude. Only the power requires pilot manual control.

Power Management

Unless the aircraft has an autothrottle system, you must adjust the power to an appropriate setting when performing any climb, descent, or level-off. You cannot allow the aircraft to exceed any applicable speed limitations during a descent. During a climb at a vertical speed that the aircraft cannot sustain, the FD/autopilot may command a pitch that results in a stall.

Essential Skills

1. Use the FD/autopilot to climb or descend to and automatically capture an assigned altitude.

2. Determine the indications of the armed or capture mode, and what pilot actions will cancel those modes.

3. Determine if the system allows resetting of the armed or capture mode, or if manual control is the only option after cancellation of these modes.

4. Determine the available methods of activating the altitude armed or capture mode.

5. Determine the average power necessary for normal climbs and descents. Practice changing the power to these settings in coordination with making the FD/autopilot mode changes.

6. Determine and record maximum climb vertical speeds and power settings for temperatures and altitudes. Ensure the values are in agreement with values in AFM/POH for conditions. Make note of highest practical pitch attitude values, conditions, and loading. Remember powerplant factors (e.g., minimum powerplant temperature, bleed air requirements) and airframe limitations (e.g., VA in setting power).

Course Intercepts

Flying an Assigned Heading To Intercept a Course or VOR Radial

You can use the navigation mode in combination with the heading function to fly an assigned heading to intercept a course. The procedure illustrated in *Figure 4-13* takes advantage of the ability to arm the navigation mode while the heading mode is engaged.

Figure 4-13 illustrates selecting the assigned heading, setting up your FD/FMS autopilot for the assigned course, engaging the heading mode, and arming the navigation function. Once the aircraft reaches the course, the autopilot automatically disengages the heading function and engages the navigation mode.

On most FD/autopilots, courses can be intercepted by first using the heading "bug" to select an intercept course and then engaging the heading function. Alternatively, engaging the navigation function in some units causes the FD/autopilot to select an intercept heading, engage the heading function, and arm the navigation function. This can be a cause for conflict if ATC assigns an intercept heading, but the FD is programmed to use one angle. In those instances, you need to set the heading into the FD/autopilot, fly, and control the intercept until the aircraft is close enough to complete the intercept and capture without deviating from the ATC instructions. At that point you can select and arm the navigation mode, which completes the intercept and begins tracking the selected course.

Essential Skills

1. Use the FD/autopilot to fly an assigned heading to capture and track a VOR and/or RNAV course.

2. Determine if the FD/autopilot uses preprogrammed intercepts or set headings for navigation course interceptions.

3. Determine the indications of navigation mode armed conditions.

4. Determine parameters of preprogrammed intercept modes, if applicable.

5. Determine minimum and maximum intercept angle limitations, if any.

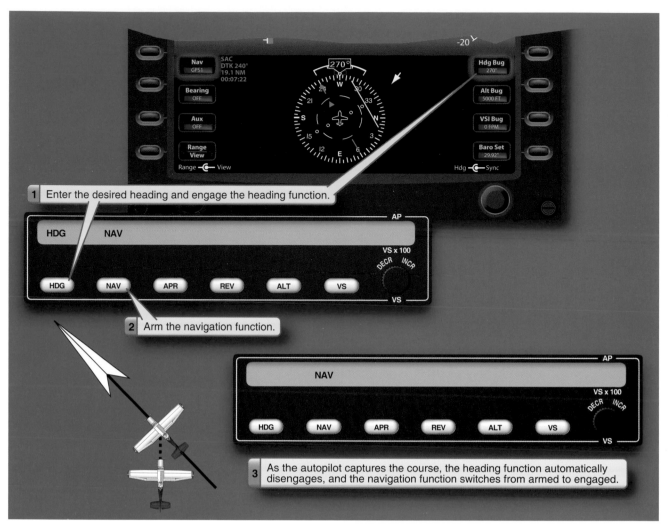

Figure 4-13. *Flying an assigned heading to intercept a course.*

Coupled Approaches

The approach function is similar to the navigation mode, but flies the selected course with the higher degree of precision necessary for instrument approaches and allows glideslope tracking in the vertical dimension. Most autopilots feature a separate button that allows you to engage the approach function, as shown in *Figure 4-14*. (NOTE: Usually, this mode is not used with most GPS receivers. The GPS approach RNP (required navigation performance) of 0.3 induces the necessary flight tracking precision. This mode is used only if specifically stated as a command in the avionics handbook for that equipment in that aircraft).

Like the navigation function, the approach mode can be used to execute precision and nonprecision approaches that rely on types of ground-based navigation facilities (e.g., VOR, VOR/DME, and localizer approaches).

ILS Approaches

Coupled ILS approaches make use of the autopilot's glideslope function. *Figure 4-15* shows the procedure for one type of autopilot.

Note that you cannot directly arm or engage the glideslope function. The autopilot must usually be engaged first in the approach and altitude modes. When the FD/autopilot begins to sense the glideslope, the glideslope function will automatically arm. When the aircraft intercepts the glideslope, the glideslope function engages automatically, and uses the aircraft's pitch control to remain on the glideslope. It is important to note that, generally, the glideslope function can capture the glideslope only from below or on glideslope.

RNAV Approaches With Vertical Guidance

Coupled RNAV approaches with vertical guidance work in the same way as coupled ILS approaches. Lateral and vertical guidance commands are generated by the FMS/NAV and sent to the FD/autopilot. The same approach and glideslope

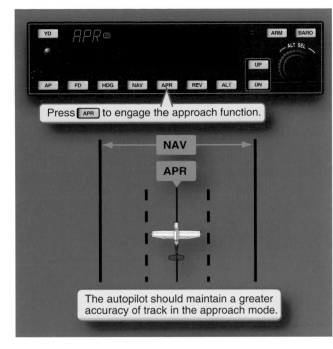

Press [APR] to engage the approach function.

The autopilot should maintain a greater accuracy of track in the approach mode.

Figure 4-14. *Flying a coupled nonprecision approach.*

functions of the autopilot are used in the same way to carry out the lateral and vertical guidance and control of the aircraft. This process is transparent to the pilot. Most "VNAV" functions do not qualify as approach vertical functions and many FMS/GPS units inhibit that function during approaches.

Power Management

Since most autopilots are not capable of manipulating power settings, you must manage the throttle to control airspeed throughout all phases of the approach. The power changes needed during altitude changes must supply the necessary thrust to overcome the drag. The pilot must coordinate the powerplant settings with the commands given to the FD/autopilot. Remember, the FD/autopilot can control the aircraft's pitch attitude only for altitude or airspeed, but not both. The FD/autopilot attempts to perform as programmed by you, the pilot. If the climbing vertical speed selection is too great, the aircraft increases the pitch attitude until it achieves that vertical speed, or the wing stalls. Selection of an airspeed or descent rate that is too great for the power selected can result in speeds beyond the airframe limitations. Leveling off from a descent, without restoring a cruise power setting results in a stall as the FD/autopilot attempts to hold the altitude selected.

Essential Skills

1. Use the FD/autopilot to couple to a precision approach.

2. Use the FD/autopilot to couple to a nonprecision approach.

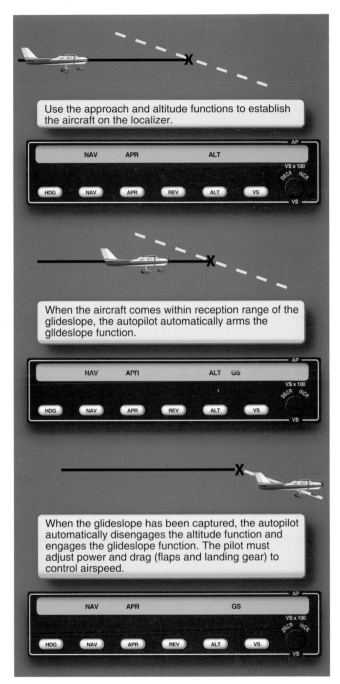

Use the approach and altitude functions to establish the aircraft on the localizer.

When the aircraft comes within reception range of the glideslope, the autopilot automatically arms the glideslope function.

When the glideslope has been captured, the autopilot automatically disengages the altitude function and engages the glideslope function. The pilot must adjust power and drag (flaps and landing gear) to control airspeed.

Figure 4-15. *Flying a coupled precision approach.*

3. Use the FD/autopilot to couple to an RNAV approach.

4. Determine the power setting required to fly the approaches.

5. Determine the power settings necessary for leveloff during nonprecision approaches and go-around power settings for both precision and nonprecision approaches.

6. Determine the speeds available for the minimum recommended powerplant settings. It is useful to determine if an ATC clearance can be accepted for climbs, altitudes, and descents.

Deciding When To Use the FD/Autopilot

In addition to learning how to use the FD/autopilot, you must also learn when to use it. Since there are no definitive rules about when an FD/autopilot should or should not be used, you must learn to consider the benefits and disadvantages of using the FD/autopilot in any given situation.

One of the most valuable benefits of using the FD/autopilot is delegating the constant task of manipulating the aircraft's controls to the equipment, which do nothing other than comply with the pilot's programming. This allows you more time to manage and observe the entire flight situation. Managing the flight versus actually moving the controls allows more time for:

1. Programming. Especially when flying under IFR, changes to a route are inevitable. Even when the pilot is proficient in using FMS/RNAV, this task requires focusing some attention on the programming task. The FD/autopilot keeps the aircraft on the programmed heading or course and altitude while the pilot makes the necessary changes to the flight plan. If programmed correctly, the aircraft maintains the correct track and altitude.

2. Distracting tasks/workload. Similarly, the FD/autopilot is used to control basic aircraft movement while the pilot focuses attention on tasks such as reviewing charts, briefing and configuring for an instrument approach, updating weather information, etc. The FD/autopilot can also be a great help in other high workload situations, such as flying in a busy terminal area or executing a missed approach in adverse weather conditions.

3. Maintaining autopilot skills. The FD/autopilot's ability to help manage pilot workload depends heavily on the pilot's proficiency in using it. Regular practice with the various autopilot functions (especially the approach functions) is essential to develop and maintain the knowledge and skills necessary to maximize its utilization.

4. Emergencies. The FD/autopilot can be extremely useful during an emergency. It can reduce pilot workload and facilitate efforts to troubleshoot the emergency.

Disadvantages of using the FD/autopilot include the following:

1. Forgetting to maintain manual flying skills. It is important to practice flying without the FD/autopilot often enough to maintain proficiency in basic flying skills and the instrument cross-check and scan. One common pitfall of advanced avionics is the pilot's tendency to forget to maintain hard-earned skills for instrument flight. All equipment will fail at some time. The competent pilot is ready and prepared to make a transition to aircraft piloting at any time.

2. Turbulence. The pilot's operating handbook (POH) and FD/autopilot flight manual supplements for many aircraft discourage or prohibit use of the autopilot's altitude hold function during moderate or severe turbulence. Some FD/autopilot systems may default or disengage if certain trim or control limits are encountered during turbulent conditions. You should consult the flight manual to ensure the aircraft is not operated outside specified limits. The aircraft's flightpath and mode indications should always be monitored to ensure what modes are active.

3. Minimum altitude. Autopilots are certified for use above a specified minimum altitude above ground level (AGL). Some higher performance and higher service ceiling aircraft require autopilot control above certain airspeeds and altitudes. The flight manual and operations manual (if any) should be consulted to ensure that the pilot does not operate the aircraft outside specified limits. For higher safety standards, commercial operators must observe restrictions in Title 14 of the Code of Federal Regulations (14 CFR) sections 121.579, 125.328, and 135.93, according to their regulatory classification. Adoption of these limits by private operators would add a safety margin to flights conducted under 14 CFR part 91 regulations in many cases.

4. Possible malfunction. If at any time the pilot observes unexpected or uncommanded behavior from the autopilot, he or she should disengage the autopilot until determination of the cause and its resolution. Most autopilot systems have multiple methods of disengagement; you should be immediately aware of all of them. Also be aware of the methods to cancel the FD display to avoid confusing information.

Miscellaneous Autopilot Topics

Autopilot Mode Awareness

In addition to performing the basic aircraft control and navigation function described previously, some autopilots are capable of automatically switching from one function to another. These automatic mode changes can complicate the task of maintaining mode awareness, but every autopilot has some form of flight mode annunciator that shows which autopilot functions are currently engaged. The autopilot shown in *Figure 4-4* displays the name of any autopilot mode that is currently engaged just above the button used to engage the function. It is important to develop two habits:

1. Checking the flight mode annunciator after entering a command to ensure that the selected function is indeed armed or engaged, as appropriate.

2. Including the flight mode annunciator in the scan to maintain continuous awareness of what mode is active and what is armed to activate next.

Positive Exchange of Controls

When control of the aircraft is transferred between two pilots, it is important to acknowledge this exchange verbally. The pilot relinquishing control of the aircraft should state, "You have the flight controls." The pilot assuming control of the aircraft should state, "I have the flight controls," and then the pilot relinquishing control should restate, "You have the flight controls." Following these procedures reduces the possibility of confusion about who is flying the aircraft at any given time.

Using an FD/autopilot system can present an opportunity for confusion. When engaging the autopilot, it is a good idea to announce that the autopilot is being engaged, what autopilot mode is being used, and then to confirm the settings using the flight mode annunciator. It has been general practice for many years in many aircraft to first engage the FD to determine what instructions it was going to transmit to the autopilot. This is determined by reading the FD's command bars. If the commands shown agree with your perception of the control motions to be made, then engage the autopilot to fly the entered course and vertical mode. A caution at this point: some FDs cancel the altitude hold mode when the autopilot is engaged. Always ensure that, after autopilot engagement, the desired modes are still active.

Preflighting the Autopilot

The POH or aircraft flight manual (AFM) supplement for each FD/autopilot system contains a preflight check procedure that must be performed before departure. As with other preflight inspection items, this check allows you to ensure that the autopilot is operating correctly, before depending on it in the air.

Autopilot and Electric Trim System Failures

It is vital that you become immediately familiar with the procedures required to disconnect or disable the electric trim and autopilot systems. Electric trim and autopilot failures can occur in the form of failure indications; unusual, unexpected, or missing actions; or, in the extreme case, a runaway servo actuator in the autopilot or trim system.

The first and closest method of disconnecting a malfunctioning autopilot is the autopilot disconnect switch, typically mounted on the control yoke. This switch is usually a red button, often mistaken by new pilots for the radio transmit button. You need to know which buttons activate which functions.

Most systems may be disconnected by the mode buttons on the autopilot control panel. However, there are some failures (shorted relays, wires, etc.) that remove control of the servo actuator from the control unit itself. In those rare instances, the pilot must find and pull the circuit breakers that interrupt power to both the trim and autopilot systems. Some trim systems have separate circuit breakers for trim motors that operate different control surfaces (roll, pitch, yaw). Many pilots have installed small plastic collars on the autopilot to facilitate finding and pulling the correct autopilot circuit breaker to kill the power to that circuit. Ensure that you understand all functions and equipment are lost if those (and in fact, any circuit breaker) are disabled. In too many cases, a circuit breaker installed in an aircraft supplies power to more functions than the label implies. To be absolutely sure, check the wiring diagrams, and do not pull circuit breakers unless the POH/AFM directs that specific action.

Another method of maintaining flight control when faced with a failed trim or autopilot system is the control yoke. Most autopilot and trim systems use a simple clutch mechanism that allows you to overpower the system by forcing the control yoke in the desired direction. This is usually checked during the afterstart/pretakeoff/ runup check.

Essential Skills

1. Demonstrate the proper preflight and ground check of the FD/autopilot system.

2. Demonstrate all methods used to disengage and disconnect an autopilot.

3. Demonstrate how to select the different modes and explain what each mode is designed to do and when it will become active.

4. Explain the flight director (FD) indications and autopilot annunciators, and how the dimming function is controlled.

Chapter Summary

Automated flight control can make a long flight easy for you by relieving you of the tedious second-by-second manipulation and control of the aircraft. Overdependence on automated flight controls can cost you hard-earned aircraft handling skills, and allow you to lose the situational awareness important to safe flight. You must practice your skills and cross-check.

Automated flight controls require you to study and learn the system's programming and mode selection actions. You must also learn what actions disconnect the autopilot, whether commanded or not. In preflight planning you must determine the limitations on the autopilot and what the installation in that aircraft permits.

It is important for you to be aware of what functions are automated and what activates those functions, and the actions or conditions that cancel or inhibit those functions. Remember that, in most aircraft, you must set the power and manage the powerplant(s). Even in very expensive aircraft equipped with autothrottle, you must monitor the powerplant(s) and be ready to intervene to ensure operation within safe parameters.

Chapter 5
Information Systems

Introduction

This chapter introduces information systems available in the advanced avionics cockpit. These systems support you in following flight progress, and in avoiding terrain, traffic, and weather hazards en route. A moving map continuously displays the aircraft's position relative to the intended route of flight, and helps you maintain the "big picture" (situational awareness) as your flight progresses. A Terrain Awareness and Warning System (TAWS) color codes surrounding terrain to make it easily apparent when terrain poses a threat. Weather systems provide in-flight access to many of the same weather products available on the ground. A fuel management system makes predictions about fuel remaining at each waypoint along the route, and helps monitor actual fuel use as your flight progresses.

Since the volume of information now available in the cockpit cannot be presented on a single display, or would clutter a single display to the point of unintelligibility, you must decide what information is needed at any given point in your flight. You will learn how information systems can be used to enhance situational awareness and increase the safety margin. It is important to avoid the pitfalls of using enhanced weather, traffic, and terrain information to fly closer to hazardous situations. This negates any safety advantage created by advanced avionics. Advanced avionics aircraft crash due to the same causes as aircraft with traditional instrumentation.

Multi-Function Display

A multi-function display (MFD) presents information drawn from a variety of aircraft information systems. Many installations allow overlay or inclusion of systems indications in the primary flight display (PFD), in addition to the primary flight instrument indications. By allowing you to view information provided by any one or combination of the installed systems, the MFD prevents the need for a separate display for each system. As does the flight management system (FMS)/area navigation (RNAV) unit, the MFD allows you to select different pages from different chapters that display information provided by various aircraft systems. Other controls allow you to combine information from multiple systems on one page. In many installations, the MFD serves as the backup or faildown display for the PFD. You should have a working knowledge of the faildown selection procedures and how to select the necessary data displays for the current flight phase. The controls provided by one MFD are shown in *Figure 5-1*.

Invest the time required to become a skilled user of the MFD. Familiarity with the capabilities of the MFD not only increases the information available, but also allows quick access to that information for safe flight decision-making.

Essential Skills

1. Program the multi-function display to show data provided by any aircraft system.

2. Determine how many data displays can be combined in one display.

3. Know how to select the PFD displays on the MFD, if available.

4. Determine which data displays can be overlaid onto the PFD as well as the MFD.

Moving Maps

The moving map function uses the MFD to provide a pictorial view of the present position of the aircraft, the route programmed into the FMS, the surrounding airspace, and geographical features. Moving maps offer a number of options that allow you to specify what information is presented on the MFD and how it is displayed. Moving maps typically offer several different map orientations (e.g., north up, track up), a range control that allows you to "zoom" in and out to see different volumes of airspace, and a means to adjust the amount of detail shown on the display (declutter). The moving map display does not replace looking outside the aircraft to avoid other aircraft and obstructions.

Using the Moving Map

A moving map display has a variety of uses that can aid your awareness of position and surroundings during almost any phase of flight. Verification of the displayed data with a chart accomplishes three functions:

1. Provides you practice for retention of your map reading skills.

2. Contributes to your readiness for continued safe navigation to a destination in the event of equipment problems.

3. Ensures that you maintain situational awareness.

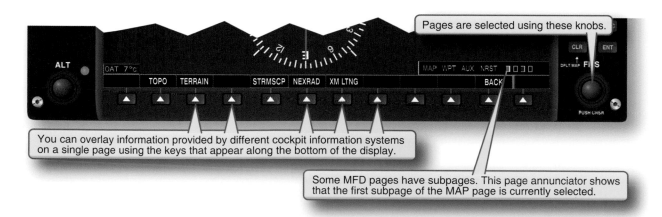

Figure 5-1. *Selecting pages on an MFD.*

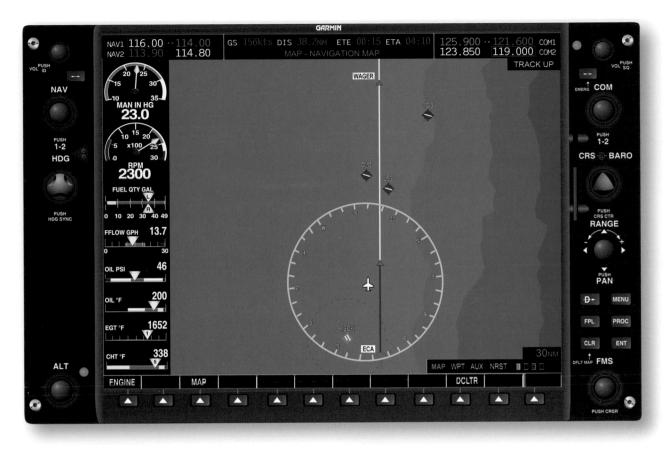

Figure 5-2. *A moving map provides the "big picture."*

Maintaining the "Big Picture"

Moving map displays can help you verify a basic understanding of the planned route and aircraft position with respect to the route, nearby terrain, and upcoming waypoints. For example, the moving map display in *Figure 5-2* shows the aircraft slightly to the left of the programmed flight route, presumably heading in the correct direction, and operating to the west of rising terrain.

Maintaining Awareness of Potential Landing Sites

The moving map in *Figure 5-2* makes several nearby alternative airports visually apparent. A classic technique used by pilots to maintain awareness is to periodically ask the question, "Where would I go if I lost engine power?" The moving map can be used in this way to maintain preparedness for an emergency, if you are aware of the map scale and aircraft capabilities.

Maintaining Awareness on the Airport Surface

On most units, you can change the range on the moving map to see a more detailed picture of the airport surface while operating on the ground. This feature is especially useful when the arrangement of runways and taxiways is complex. The moving map in *Figure 5-3* shows the aircraft prepared to taxi onto one of two possible runways.

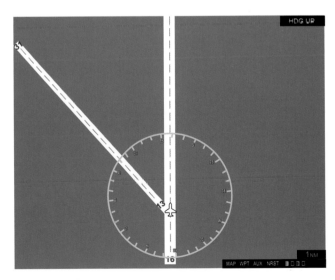

Figure 5-3. *Using the moving map on the airport surface.*

Identifying Controlled Airspace

Most moving map displays can portray surrounding airspace as well as the vertical limits of each airspace segment. This feature is particularly useful during visual flight rules (VFR) flights, but can also serve to remind you of speed restrictions that apply to airspace transitions during instrument flight rules (IFR) flight.

Identifying the Missed Approach Point

The moving map display is an especially useful aid for recognizing arrival at various points, including the missed approach point during an instrument approach. The moving map display complements the distance readout on the PFD/MFD/FMS. *Figure 5-4* shows two indications of an aircraft arriving at a missed approach point. The position of the aircraft on the moving map is very clear, and the range setting has been used to provide a more detailed view of the missed approach waypoint.

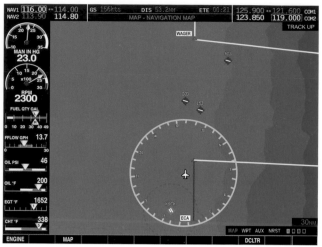

Figure 5-5. *Moving maps help make route programming errors evident.*

pictorially. For this reason, a display such as a moving map is sometimes referred to as an error-evident display. The PFD selected track indicates incorrect settings. Always be ready and able to fly the aircraft according to any air traffic control (ATC) clearance or instructions. Disengaging all automation and then reestablishing heading, track, and altitude control is the pilot's first priority at all times. Then, when the aircraft is on an assigned track at a safe altitude, pilot time can be expended to reprogram as necessary.

Catching Errors: Using The Moving Map To Detect Configuration Errors

Moving maps can help you discover errors made in programming the FMS/RNAV and PFD. The moving map display shown in *Figure 5-6* removes the depiction of the leg to the active waypoint when the FMS/RNAV is engaged in the nonsequencing mode. This feature provides an easy way to detect the common error of forgetting to set the computer back to the sequencing mode.

Figure 5-4. *The missed approach point shown on two different displays.*

CAUTION: Some units can be set to change ranges automatically. In some instances, this can lead to a loss of situational awareness as you forget or miss a scale change. This can lead to sudden pilot realization at some point that the aircraft is too high, too far, or moving too fast. Manual switching (pilot selection) of the range display ensures that you are constantly aware of the distances and closure rates to points.

Catching Errors: Using the Moving Map to Detect Route Programming Errors

Moving maps are particularly useful for catching errors made while entering modifications to the programmed route during flight. Misspelled waypoints are often difficult to detect among a list of waypoints. The moving map in *Figure 5-5* shows a route containing a misspelled waypoint. It is easy to detect the mistake when the information is shown

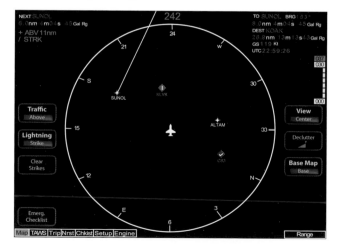

Figure 5-6. *A reminder that the FMS/RNAV is set in the nonsequencing mode.*

The moving map shown in *Figure 5-7* allows you to discover a more serious programming error quickly. In this situation, the pilot is attempting an RNAV approach. However, the course deviation indicator (CDI) has erroneously been set to display very high frequency (VHF) omnidirectional range (VOR) course indications. The CDI suggests that the aircraft is well to the west of course. The moving map display shows the true situation—the aircraft is on the RNAV approach course, but is about to depart it.

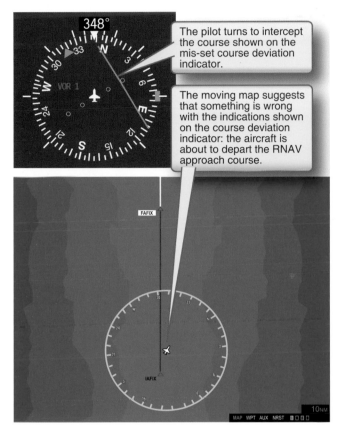

Figure 5-7. *Discovering an incorrect navigation source selection using the moving map.*

Maintaining Proficiency: Spatial Reasoning Skills

Consider the CDI shown in *Figure 5-8*. What is the position of the aircraft with respect to the VOR station? Interpreting this type of display requires more effort than interpreting the moving map, which automatically displays the solution to the position-finding problem. Pilots should expend the effort to practice this skill set. Those who learn to navigate using ground-based radio navigation aids are forced to develop spatial reasoning and visualization skills, but a Federal Aviation Administration (FAA) study showed that this type of skill tends to fade quickly when not used. Be sure to keep your spatial reasoning and visualization skills sharp.

Figure 5-8. *Can you determine your position from a simple CDI?*

Failure Indications

Failure indications on the moving map can be quite subtle. The MFD in *Figure 5-9* reflects a loss of position information, indicated by the removal of the aircraft symbol, compass labels, and other subtle differences. Be familiar with the failure indications specific to your equipment.

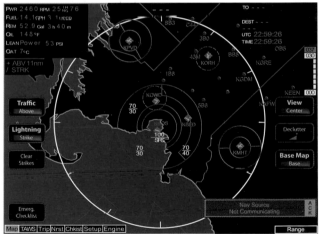

Figure 5-9. *An MFD indicating a loss of position information.*

Common Error: Using the Moving Map as a Primary Navigation Instrument

The rich detail offered by the moving map display invites you to use the display as a primary navigation instrument, but you need to resist this temptation. The moving map display is designed to provide supplemental navigation information, but is not approved as a substitute for primary navigation instruments. The moving map is not required to meet any certification standards for accuracy or information as are the primary navigation CDI and related system components. Bear in mind that the apparent accuracy of the moving map display can be affected by factors as simple as the range setting of the display. An aircraft 10 miles off course can appear to be centered on an airway when the range is set to cover great distances.

Awareness: Overreliance on the Moving Map

With the position of the aircraft conveniently displayed at all times on a color screen in front of you, it is easy to let the computers do the work of monitoring flight progress. Numerous studies have demonstrated that pilots have a tendency to monitor and process navigational information from conventional sources (e.g., outside reference or conventional navigation instruments) much less actively when a moving map display is available. In a National Aeronautics and Space Administration (NASA) study, two groups of pilots were asked to navigate along a circuit of checkpoints during a VFR cross-country flight. One group navigated using a sectional chart and pilotage. The other group had the same sectional chart plus an RNAV computer and a moving map. After completing the circuit, both groups were asked to navigate the circuit again, this time with no navigational resources. Pilots who had navigated with only the sectional chart performed well, finding the checkpoints again with reasonable accuracy. The performance was less favorable by pilots who had the FMS/RNAV and moving map available. While half of these pilots found the checkpoints with reasonable accuracy, one-fourth of the pilots made larger errors in identifying the checkpoints. The remaining pilots were wholly unable to find their way back to the airport of origin. This study makes two important points:

1. The existence of information about aircraft position and geographical surroundings in an FMS/RNAV and moving map display does not mean that the pilot maintains true situational awareness or involvement with the operation of the flight to a degree needed for a safe outcome.

2. The key to the successful use of a moving map display is to use the display as a supplement—not a substitute—for active involvement in the navigational process.

What does it take to use a moving map and remain "in the loop," or situationally aware? In a second NASA study, pilots who used an FMS/RNAV and moving map display were asked to act as "tour guides," pointing out geographical features to a passenger while navigating the same set of cross-country checkpoints. When confronted with a surprise request to navigate around the circuit again with the FMS/RNAV and map turned off, these pilots performed as well as anyone else. The simple task of pointing out geographical features was enough to avoid the out-of-the-loop phenomenon.

A moving map provides a wealth of information about your route of flight and gives you the opportunity to consider many similar questions along the way. Where would you land if you lost engine power? Which alternate airport would you use if weather at your destination deteriorated below minimums? Which nearby VOR stations could be used (and should be tuned as the flight progresses) in the event that the global positioning system (GPS) signal or other RNAV navigation data source is lost? Is a more direct routing possible? Diligent pilots continually ask questions like these.

Terrain Systems

Terrain systems provide information about significant terrain along your route of flight. Terrain systems were designed to help reduce controlled flight into terrain (CFIT) accidents. Remember, however, that use of these terrain proximity information systems for primary terrain avoidance is prohibited. The terrain proximity map is intended only to enhance situational awareness; it remains the pilot's responsibility to ensure terrain avoidance at all times. Safe flight practices include pilot knowledgeability of the maximum elevation figures (MEF), published in blue for each grid square on sectional charts, and planning flight altitudes above those elevations. Despite all efforts by the charting agency to be current, there will always be obstructions in place before the documentation arrives for charting. Therefore, the competent pilot always allows for sufficient clearance for unknown towers and buildings. Experienced pilots have learned that many aircraft cannot outclimb certain mountainous slopes. You should always fly down (descend) into a valley or canyon, rather than attempting to fly up the valley and become trapped in a box canyon too narrow for a turn and too steep to climb over. One regularly overlooked factor is the loss of power generally associated with the higher elevations at which canyons and steep slopes are often found.

Early Systems

Various terrain avoidance systems have been certificated and used in the past. One early system was termed Ground Proximity Warning System (GPWS, often pronounced "GipWhiz"). One major shortcoming of the system was a lack of predictive terrain warnings. Most early systems simply used a radar altimeter as the sensor. The radar altimeter simply indicated the altitude of the aircraft above the ground immediately below the airframe. The subsequently developed enhanced GPWS (EGPWS or eGPWS) used GPS location data combined with a worldwide terrain database to predict that a canyon wall was just ahead and a climb should be started. The older GPWS had no indication of a very close hazard. However, the system did prevent numerous gear-up landings and offered warnings when terrain presented a slope to much higher terrain.

Terrain Display

The most basic type (not necessarily certified) of terrain system is the terrain display. A terrain display uses the MFD to plot the position of the aircraft against a pictorial presentation of surrounding terrain. A terrain display usually relies on a GPS location signal to compare the position and altitude of the aircraft against the terrain found in an internal topographical database. *Figure 5-10* shows the position of the aircraft and surrounding terrain displayed on an MFD.

Terrain displays use a simple color-coding convention to portray the difference between the present altitude of the aircraft and the height of the surrounding terrain. Terrain more than 1,000 feet below the aircraft is coded black. Terrain less than 1,000 feet but more than 100 feet below the aircraft is coded yellow. Terrain less than 100 feet below the aircraft is coded red. Man-made obstacles (e.g., radio towers, power lines, buildings) generally do not appear in a topographical database.

Monitoring Surrounding Terrain During Departure and Arrival

Terrain displays are especially useful during departure and arrival phases of flight. For example, the aircraft shown in *Figure 5-10* has departed Denver and is heading for a waypoint situated in high terrain in the Rocky Mountains. A pilot with good situational awareness has many concerns during the departure. For instance, is climb performance meeting expectations? The terrain display reduces the need for you to perform mental calculations by verifying that the depiction of the terrain that lies ahead of the aircraft is steadily changing from red to yellow to black. If the terrain depiction remains red as the aircraft approaches, you know there is a problem. Similarly, if the aircraft has been assigned a heading and altitude vector by air traffic control, the terrain display provides a simple way of monitoring the safety of these directives. If the equipment is not certified as meeting TAWS requirements (see Title 14 of the Code of Federal Regulations (14 CFR) part 91, section 91.223), the accuracy can be in doubt.

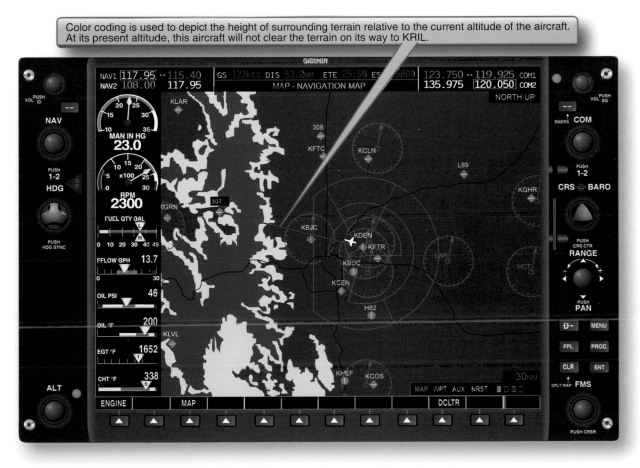

Figure 5-10. *Terrain depicted on an MFD.*

Evaluating a Direct-To Routing

One of the risks involved in proceeding directly to a waypoint is that you may be yet unaware of any significant terrain between the present position and the waypoint. A terrain display offers a convenient way of seeing clearly what lies between here and there as reported and documented in the database. Always consult the MEF values along the path of flight to ensure terrain and obstruction clearance.

Terrain Awareness and Warning Systems

A terrain awareness and warning system (TAWS) offers you all of the features of a terrain display along with a sophisticated warning system that alerts you to potential threats posed by surrounding terrain. A terrain awareness and warning system uses the aircraft's GPS navigation signal and altimetry systems to compare the position and trajectory of the aircraft against a more detailed terrain and obstacle database. This database attempts to detail every obstruction that could pose a threat to an aircraft in flight.

TAWS A and TAWS B

There are presently two classes of certified terrain awareness and warning systems that differ in the capabilities they provide to the pilot: TAWS A and TAWS B.

A TAWS A system provides indications for the following potentially hazardous situations:

1. Excessive rate of descent

2. Excessive closure rate to terrain

3. Altitude loss after takeoff

4. Negative climb rate

5. Flight into terrain when not in landing configuration

6. Excessive downward deviation from glideslope

7. Premature descent

8. Terrain along future portions of the intended flight route

A TAWS B system provides indications of imminent contact with the ground in three potentially hazardous situations:

1. Excessive rate of descent

2. Excessive closure rate to terrain (per Advisory Circular (AC) 23-18, to 500 feet above terrain)

3. Negative climb rate or altitude loss after takeoff

TAWS Alerts

Aural alerts issued by a terrain awareness and warning system warn you about specific situations that present a terrain collision hazard. Using a predictive "look ahead" function based on the aircraft's ground speed, the terrain system alerts you to upcoming terrain. At a closure time of approximately 1 minute, a "Caution! Terrain!" alert is issued. This alert changes to the more serious "Terrain! Terrain!" alert when the closure time reaches 30 seconds. In some areas of the world, this terrain warning may very well be too late, depending on the performance of the aircraft. You need to determine the equipment's criteria and note if the unit makes allowances for lower power output of the powerplant(s) at higher elevations, resulting in lower climb rates than may be programmed into the unit for that aircraft.

A second type of aural alert warns about excessive descent rates sensed by the system ("Sink Rate!") or inadvertent loss of altitude after takeoff ("Don't Sink!").

The introduction of terrain awareness and warning systems has sharply reduced the number of CFIT accidents. Despite this significant leap forward in safety, incidents and accidents involving terrain still happen. In the modern TAWS-equipped cockpit, some of these incidents have been related to pilot reaction to TAWS alerts. TAWS sometimes gives nuisance alerts that desensitize the pilot to TAWS alerts, which can result in the pilot's decision to ignore a valid alert deemed unnecessary by the pilot. Most TAWS systems contain software logic that attempts to recognize and remain silent in situations in which proximity to terrain is normal. This logic is partly based on the aircraft's distance from the runway of intended landing. For example, flying at an altitude of 200 feet AGL when 3,500 feet away from the runway is reasonable, but flying at an altitude of 200 feet AGL when 5 miles from the runway is not reasonable. TAWS' logic attempts to silence itself in normal situations, and to sound in abnormal situations.

Risk: Silencing TAWS Alerts

Despite efforts to minimize nuisance alerts, they still occur occasionally. For this reason, most TAWS systems offer a terrain inhibit switch that allows you to silence TAWS alerts. There have been cases in which pilots have used the inhibit switch or ignored TAWS alerts, thinking they were nuisance alerts, when in fact the alerts were valid indications of a dangerous situation. For this reason, you should train yourself to respond to TAWS alerts just as you would to any other sort of emergency. Always, if in any doubt, set "Full Power and Climb" at V_X or V_Y, depending on the equipment manual and AFM/POH. The practice of simply ignoring or disabling TAWS alerts based on pilot intuition has not proved to be a safe one. Your manufacturer's reference manual and aircraft flight manual supplement will prescribe specific procedures for responding to TAWS alerts.

The only current, fully certified systems, known as TAWS, are certified under Technical Standards Order (TSO)-C151.

TAWS equipment is required for turbine-powered airplanes having six or more passenger seats and manufactured after certain dates (see 14 CFR part 91, section 91.223). TAWS is now an affordable option in many advanced avionics due to decreased cost and increased capabilities of computer circuits and components. All aircraft would be safer with TAWS and crews trained to use the technology.

Risk: Flying in Close Proximity to Terrain

A display that plainly shows your position with respect to surrounding terrain is sometimes cited as the most reassuring system available in the advanced avionics cockpit. The same display can also invite the unwary pilot to attempt risky maneuvers. Suppose that, on a VFR flight to an airport located in hilly terrain, you encounter a layer of fog at 1,100 feet. In an aircraft with no terrain system, you would not consider proceeding to the airport because you have a personal minimum of 1,500 feet. With a ceiling of less than 1,500 feet, you deem the situation simply too risky. With the surrounding terrain neatly displayed in front of you, however, you may feel more confident and be tempted to give it a try. However, a wise pilot remembers that, unless the equipment is TAWS certified, accuracy is suspect. Even with TAWS certification, the information presented is no better than the database accuracy. Consult the equipment handbook or manual to determine the accuracy of the database in that area.

CFIT accidents are still occurring despite the advent of advanced avionics. What has happened here? Psychologist Gerald J. S. Wilde coined the phrase risk homeostasis to refer to a tendency for humans to seek target levels of risk. Our hill-flying scenario illustrates the concept. After pondering the perceived risks, you decide that having the terrain display gives you the same level of perceived risk with a 1,000-foot ceiling as you felt you had at 1,500 feet without the terrain display. You see no need to "give away" this new margin of perceived safety. Rather, you decide to use it to your advantage. Equipped with the terrain display, your new minimum ceiling becomes 1,000 feet, and you continue on your way to the airport.

Wilde does not support the idea of using technology to seek target levels of risk. Rather, he argues that safety measures such as seat belt laws and anti-lock brakes have not resulted in drastic reductions in highway fatalities in part because, in response to the added sense of safety provided by these measures, drivers have emboldened their driving behavior to maintain existing levels of risk.

Another issue is the lack of training in the new equipment and its uses. The functions of TAWS and basically how it works have been previously described, yet there is no training program outside the military that teaches anyone to fly based on the TAWS display. It requires much precision flight training to learn the timing and skills to fly from a display depicting a myriad of data and converting that data into close and low terrain flight directions. All advanced avionics are designed to help the pilot avoid a hazard, not enable the pilot to get closer to it. TAWS is not a terrain flight following system.

Cockpit Weather Systems

Advanced avionics cockpit weather systems provide many of the same weather products available on the ground and have a variety of uses that can enhance awareness of weather that may be encountered during almost any phase of flight. Radar images, satellite weather pictures, Aviation Routine Weather Reports (METARs), terminal weather forecasts (TAFs), significant meteorological information (SIGMETs), Airmen's Meteorological Information (AIRMETs), and other products are now readily accessible at any time during flight. Weather products provided by cockpit weather systems are typically presented on an MFD. Some installations allow the overlay of this data in the PFD. You must learn the procedures required to show each kind of weather product on the MFD and/or PFD, and how to interpret each type of weather product. Know the limitations of each type of product, and the ways in which cockpit weather systems can be used to gather information and remain clear of weather hazards throughout the flight.

Thunderstorms and Precipitation

Thunderstorms and general areas of precipitation are detected through the use of radar. In the advanced avionics cockpit, radar data can come from one of two sources: an onboard weather radar system or a ground weather surveillance radar system, such as the Next Generation Radar (NEXRAD) system. Ground weather surveillance system data is transmitted to the cockpit via a broadcast (or datalink) weather service. Onboard weather radar and ground weather surveillance radar systems each offer advantages and disadvantages to the pilot. Some aircraft use a combination of both systems.

While onboard radar is real time, many downloaded radar images and other reports are delayed for some time period for various reasons. Given the nature of thunderstorms and other weather hazards, this delay could prove hazardous. You must know the true quality and age of the data.

Most MFDs are capable of presenting radar data together with aircraft position and the programmed route, as shown in *Figure 5-11*.

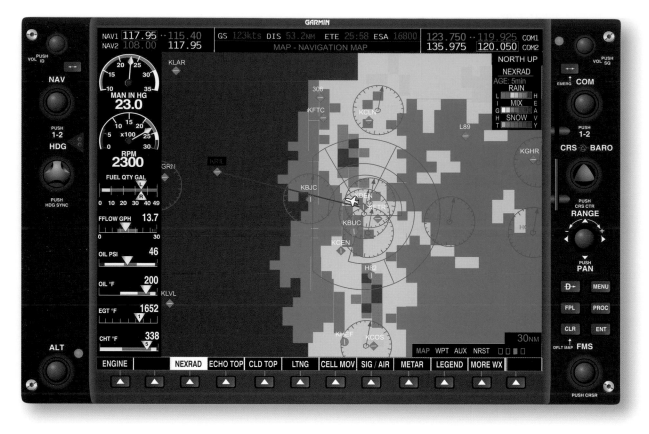

Figure 5-11. *Radar data shown on an MFD.*

Onboard Weather Radar Systems

Onboard weather radar uses an adjustable aircraft mounted radar antenna to detect, in real time, weather phenomena near the aircraft. The coverage of an onboard weather radar system is similar to a flashlight beam, as illustrated in *Figure 5-12.* You should always remember that the radar displays only areas of water or moisture (rain, sleet, snow, and hail). Radar does not display turbulence or lightning.

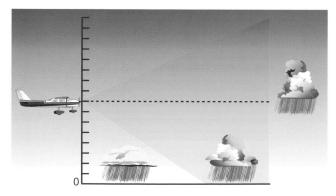

Figure 5-12. *A radar beam allows you to see some weather cells, but not others.*

Although the tilt of the radar antenna can be adjusted upward and downward, the weather phenomena that the weather radar can detect are limited in both direction and range. The radar system in *Figure 5-12* fails to detect the two cells that lie below and beyond the radar beam.

As illustrated in *Figure 5-12,* you must be careful not to assume that the only cells in the area are the ones shown on the radar display. The two additional cells in *Figure 5-12* are present, but not detected by the onboard weather radar system.

When a cell is detected by an onboard weather radar system, that cell often absorbs or reflects all of the radio signals sent out by the radar system. This phenomenon, called attenuation, prevents the radar from detecting any additional cells that might lie behind the first cell. *Figure 5-13* illustrates radar attenuation, in which one cell "shadows" another cell.

A simple color-coding scheme, as shown in *Figure 5-14,* is used to represent the intensity of radar echoes detected by an onboard weather radar system.

Ground Weather Surveillance Radar

Ground weather surveillance integrates weather information from many ground radar stations. The weather information collected from many sources is then used to create a composite picture that covers large volumes of airspace. These composite radar images can then be transmitted to aircraft equipped with weather data receivers.

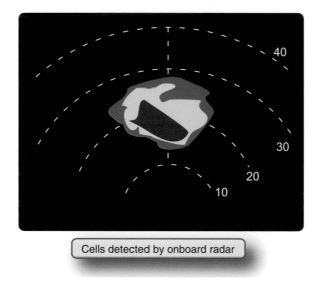

Cells detected by onboard radar

Actual situation

Figure 5-13. *Radar attenuation.*

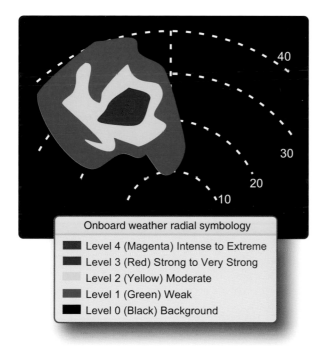

Onboard weather radial symbology

- Level 4 (Magenta) Intense to Extreme
- Level 3 (Red) Strong to Very Strong
- Level 2 (Yellow) Moderate
- Level 1 (Green) Weak
- Level 0 (Black) Background

Figure 5-14. *Color coding of intensity on an onboard weather radar system.*

Except in those areas for which no ground radar coverage is available, the range of ground weather surveillance radar systems is essentially unlimited. Ground radars have the luxury of large antennas; big, heavy power supplies; and powerful transmitters—without the constraints of aerodynamic drag, power, weight, and equipment volume restrictions and concerns.

Unlike onboard weather radar systems, weather data received from a ground weather surveillance radar system is not real-time information. The process of collecting, composing, transmitting, and receiving weather information naturally

takes time. Therefore, the radar data reflect recent rather than current weather conditions.

The color-coding scheme used by one ground weather surveillance radar system (NEXRAD) is shown in *Figure 5-15*. Note that this color-coding scheme is slightly more sophisticated than that for the onboard system in *Figure 5-13*. It is capable of distinguishing rain, snow, and mixtures of the two.

Ground weather surveillance radar symbology

Figure 5-15. *Color coding of intensity on a NEXRAD display.*

Limitations of Both Types of Weather Radar Systems

Weather radar does not detect most other kinds of hazardous weather such as fog, icing, and turbulence. The absence of radar return on a radar display does not in any way mean

5-11

"clear skies." Skillful users of weather radar are able to recover clues of other weather phenomena, such as hail and turbulence, from radar data.

A second limitation of weather radar is that the earliest (cumulus) stage of a thunderstorm is usually free of precipitation and may not be detected by radar. Convective wind shear, severe turbulence, and icing are characteristic of thunderstorms during the cumulus stage.

The pilot must beware of areas that offer no radar coverage. In many cases, these areas appear blank on a weather display. The absence of weather hazards as shown on a screen does not imply the actual absence of weather hazards.

Lightning

Most MFDs are also capable of depicting electrical activity that is indicative of lightning. Like radar data, lightning data can come from two sources: onboard and broadcast weather systems. Both systems have strengths and limitations and work together to present a more complete weather picture. Lightning data is an excellent complement to radar data for detecting the presence of thunderstorms.

An onboard lightning detection system consists of a simple antenna and processing unit that senses electrical discharges in the atmosphere and attempts to determine which electromagnetic signals have the "signature" of lightning strikes. Lightning detectors or spherics receivers, such as Stormscope® and Strikefinder ®, have been known to indicate areas of static consistent with turbulence even where there was no rain associated with the turbulence. The MFD in *Figure 5-16* depicts lightning strikes detected by a lightning detection system. Onboard lightning detection systems provide real-time information about electrical discharges. Estimates of the direction (or azimuth) of

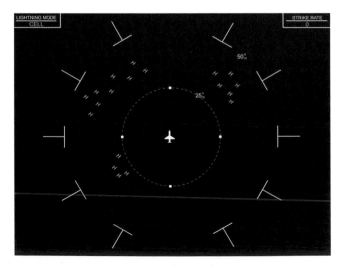

Figure 5-16. *Lightning strikes shown on an MFD.*

electrical discharges provided by an onboard lightning system are usually quite accurate. Estimates of the range (or distance) of electrical discharges tend to be less accurate.

Broadcast (or data link) weather services are also capable of transmitting lightning data to the cockpit. The symbology used to present lightning data derived from these sources is similar to that used by onboard lightning detection systems. The lightning data provided by ground weather surveillance systems is also a delayed weather product. Since the lightning data provided by a broadcast service is derived from multiple sensors, estimates of the range of electrical discharges are more accurate than those provided by onboard systems.

Clouds

Weather products that describe cloud coverage are generally available only from broadcast weather services. One popular broadcast weather service offers graphical displays of visible cloud cover along with the cloud top altitude as determined from satellite imagery. *Figure 5-17* shows an MFD that depicts cloud cover and cloud tops.

Figure 5-17. *Cloud cover and cloud tops shown on an MFD.*

Other Weather Products

Broadcast weather services offer many of the other weather products that can be obtained during a pre-flight briefing on the ground. Broadcast weather services can also provide graphical wind data, SIGMETs and AIRMETs, freezing levels, temporary flight restrictions, surface analyses, and hurricane tracks. The MFD in *Figure 5-18* shows METAR and TAF data.

Using Advanced Weather Data Systems

The increased availability of weather information is changing the way pilots think about weather briefing and the weather decision-making process. You are no longer limited to obtaining weather forecast products prior to a flight, only to discover different actual flight conditions in the air. Now

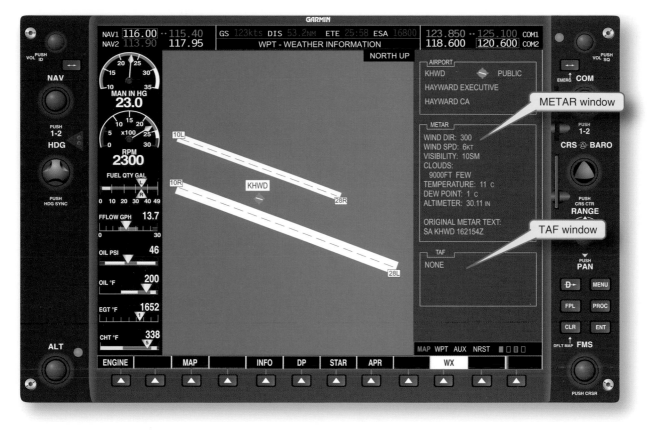

Figure 5-18. *METAR and TAF data shown on an MFD.*

more than ever, weather briefing is an activity that begins prior to departure and continues until the flight is completed.

Preflight Overview

A broadcast weather service allows you to see these products in the cockpit. Most systems offer a movable cursor that allows scrolling through the display to center on any location along the route. This capability, combined with the MFD's range control, allows you to look for significant weather anywhere along the planned route of flight prior to departure.

Track Progress of Significant Weather En Route

The same scroll and range control features allow you to look ahead and check for weather conditions along upcoming portions of the flight route. Weather forecasts, such as TAFs, SIGMETs, and AIRMETs, issued after departure can be easily checked en route.

Investigate Weather Phenomena Reported by Radio

You can use cockpit weather systems to further investigate advisories received from HIWAS and other radio broadcasts. Another practical use is to check the METAR for a destination airport before flying in range of the airport's ATIS broadcast. When you suspect that changing weather conditions have made continuation to the destination airport inadvisable, the radar and satellite features can be used to search for alternate

airports. Since not all weather products can be viewed at once, a key pilot skill is the ability to determine which weather products to display at what times.

Broadcast Weather Products Versus Onboard Weather Sensors

Onboard weather sensor systems and broadcast weather services contribute to the weather decision-making process in slightly different ways. Broadcast weather services provide delayed information over a wider coverage area. Broadcast weather services are useful for making strategic decisions about which areas to fly into and which areas to avoid. Using a broadcast weather product to attempt to find a hole in a line of thunderstorms is inappropriate, since you cannot know if the current location of the thunderstorm cells is the same as when the broadcast weather product was generated. Onboard weather sensor systems provide real-time information about weather phenomena in the immediate proximity of the aircraft. Onboard weather sensor systems are useful when making immediate, close-range decisions about flying in the vicinity of potentially hazardous weather phenomena. You must keep in mind the limitations of onboard weather systems.

Common Error: Skipping the Preflight Weather Briefing

The easy availability of weather information in the cockpit can lure you to skip the preflight weather briefing. Time pressure adds further incentive to simply jump in and go. Keep in mind that flight services stations (FSS)/automated flight service stations (AFSS) offer many advantages over an advanced weather data system, so do not use advanced avionics weather data systems as a substitute for a pre-flight weather briefing. As a simple example, when talking to an FSS/AFSS weather briefer, it is possible to get a better overall picture of the weather system and pilot reports not yet entered into the system. The FSS/AFSS briefer can also supply more Notice to Airmen (NOTAM) and other detailed information for your particular route of flight; without such briefing, the pilot might expend many precious moments searching for a critical bit of information, instead of managing the flight. Often, it is much easier to get a thorough briefing on the ground than attempting to read small reports on a bouncing MFD in a small airplane in turbulent conditions.

Traffic Data Systems

A traffic data system is designed to help you visually acquire and remain aware of nearby aircraft that pose potential collision threats. All traffic data systems provide aural alerts when the aircraft comes within a certain distance of any other detected aircraft. Traffic data systems coupled with MFDs can provide visual representations of surrounding traffic. Most traffic data systems allow you to set the sensitivity of the system and display only traffic that exists within a specific distance from the aircraft.

There are two basic types of traffic data systems available today: one using onboard sensors to detect nearby aircraft, the other relying on traffic information transmitted from ground facilities to the cockpit.

Traffic Data Systems Using Onboard Sensing Equipment

Traffic collision avoidance systems (TCAS) and traffic advisory (TA) systems use onboard sensing equipment to locate nearby aircraft and provide alerts and advisories. Both TCAS and TA systems work by querying the transponders of nearby aircraft to determine their distance, bearing, altitude, and movement relative to your aircraft. In addition, TCAS and TA systems use Mode C information from transponders to determine altitude and vertical movement of surrounding aircraft. Using these capabilities, TCAS and TA systems provide traffic alerts and advisories.

TCAS I and TA systems can issue a TA whenever another active transponder-equipped aircraft comes within an approximately 40-second range of the aircraft. Traffic advisories take the form of an aural alert: "Traffic! Traffic!"

Advanced TCAS systems (TCAS II) can also issue a resolution advisory (RA) when another active transponder-equipped aircraft comes within an approximately 25-second range of the aircraft. RAs take the form of an avoidance command that instructs you how to fly the aircraft in order to avoid the threat. An aural alert is issued that instructs the pilot to perform a vertical avoidance maneuver. Example aural alerts are: "Climb! Climb!" and "Descend! Descend!"

TCAS and traffic advisory systems use similar symbology to present traffic information. *Figure 5-19* shows four common traffic symbols used on traffic displays. The resolution advisory symbols appear only when an advanced TCAS II system is used. The colors used to display traffic symbols vary with the capabilities of the display.

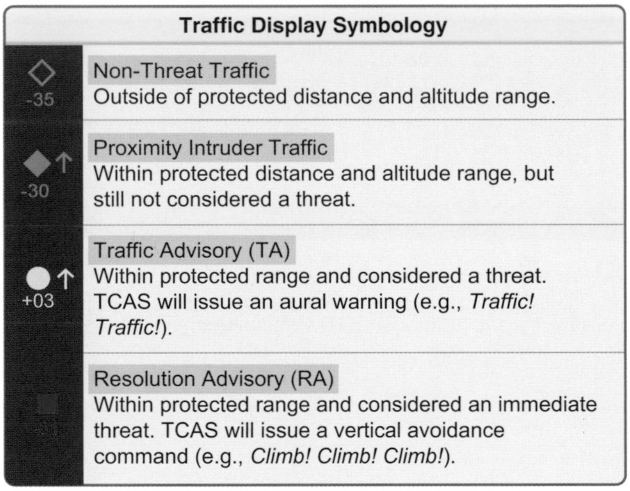

Traffic Display Symbology	
◇ -35	**Non-Threat Traffic** Outside of protected distance and altitude range.
◆↑ -30	**Proximity Intruder Traffic** Within protected distance and altitude range, but still not considered a threat.
●↑ +03	**Traffic Advisory (TA)** Within protected range and considered a threat. TCAS will issue an aural warning (e.g., *Traffic! Traffic!*).
■	**Resolution Advisory (RA)** Within protected range and considered an immediate threat. TCAS will issue a vertical avoidance command (e.g., *Climb! Climb! Climb!*).

Figure 5-19. *Traffic display symbology.*

Despite their many advantages, TCAS and traffic advisory systems have several important limitations. For example, TCAS and traffic advisory systems cannot detect aircraft that do not have active transponders. Another limitation of TCAS and traffic advisory systems is that they give unwanted alerts when the pilot is purposefully operating in the vicinity of other aircraft. For example, two aircraft making approaches to parallel runways will probably receive traffic alerts. These alerts can be distracting.

Traffic Data Systems Receiving Information From Ground-based Facilities

The traffic information service (TIS) is a second type of traffic data system. Unlike TCAS, the TIS system does not require each aircraft to have an onboard sensor that locates nearby aircraft. However, each aircraft must have operational and active transponders to be indicated on the ATC system. Rather, TIS captures traffic information that appears on radar scopes at nearby air traffic control facilities and broadcasts that information to appropriately equipped aircraft. In order to use TIS, aircraft must be equipped with a transponder capable of receiving TIS broadcasts. When TIS is operational, TIS-capable aircraft can observe traffic information in the cockpit and receive traffic advisories for proximate aircraft.

There is an important limitation of TIS. TIS data is only transmitted from approach radar facilities. No information is broadcast from en route (air route traffic control center (ARTCC)) facilities, so the effective coverage of TIS is limited to larger metropolitan areas. Some approach radar facilities are not equipped to send TIS information. Note that the aircraft must be within range (approximately 50 NM) and within line of site of the TIS station to receive broadcasts.

Advanced Traffic Data Systems Based on ADS-B

Future traffic avoidance systems will probably be able to determine position and digitally exchange information with airborne and ground-based facilities. Using the automatic dependent surveillance—broadcast (ADS-B) system, participating aircraft will continuously broadcast their own position, altitude, airspeed, trajectory, and identification to air traffic control facilities. ADS-B aircraft continuously receive the same information from like equipped aircraft in the area (line of sight), which allows onboard displays for surrounding traffic. ADS-B has been used with much success in Alaskan trials and requires less infrastructure to be usable. ADS-B equipment is demonstrating promise for better traffic separation on transoceanic routes well out of range of land-based systems. ADS-B signals are transmitted on the 978 MHz channel. The information gathered from all participating aircraft can then be transmitted back to each TIS aircraft to provide a detailed picture of the traffic situation, even if those aircraft do not have ADS-B onboard.

Using a Traffic Data System

Setting Sensitivity on a Traffic Data System

Most traffic data systems allow you to adjust sensitivity and configure the system to track targets occurring only within a specified distance and altitude. More sophisticated traffic data systems automatically adjust sensitivity throughout different phases of flight. It is important to become familiar with the use of these controls and features.

Responding to Traffic Alerts

You must develop skill in the task of visually acquiring aircraft identified by an advanced avionics traffic data system. This task requires you to use angles and distances displayed on a traffic display to help guide your visual search out the window. Since both the directions and altitudes of intruding aircraft provided to the aircraft's traffic data receiver are subject to error, you must widen your scan to all areas around the location presented on the traffic display.

When responding to air traffic control requests to acquire and maintain visual separation from nearby targets, be careful not to acknowledge contact with targets that have been observed only on a traffic display. Do not report having the traffic "in sight" before visual acquisition of a target.

Error: Overreliance on Traffic Data System/Failure To Scan

Keep in mind that a traffic data system is designed to enhance "see and avoid" capabilities. You must avoid any tendency to rely too much on traffic data systems, or to use traffic data systems as a substitute for visual scanning and acquisition of surrounding traffic. Because of the limitations of advanced traffic data systems, think of them as supplemental to your traffic awareness while you continue to assume primary responsibility to see and avoid other aircraft. Remember, too, that systems can and do fail. Traffic data systems are quite complex and any failure from the other aircraft's transponder/GPS/encoder to your receiver/decoder/GPS/traffic computer and display encoder will reduce data on your display. Power spikes, weather (lightning), and other onboard aircraft disturbances are all unknown and changing. As experience is gained with the system, better designs will eliminate problems yet to be discovered.

Using a Traffic Data System on the Ground

Most traffic data systems automatically switch to a standby mode when the aircraft is on the ground or operating below a minimum speed. The same systems typically allow you to override this feature and manually activate the traffic data system at any time. There are a number of ways to exploit this capability. When departing from an uncontrolled airport, the traffic data system can help you learn of other traffic in the vicinity of the airport. When operating in low-visibility conditions, the same feature can help inform of other aircraft operating on the airport surface. One potential future application of ADS-B is allowing controllers and pilots to monitor aircraft better on taxiways and runways. You must check equipment documentation to determine when the transponder actually transmits, where the selection controls are located, and how to use those controls.

Fuel Management Systems

A fuel management system can help you make the fuel calculations needed for in-flight decisions about potential routing, fuel stops, and diversions. A fuel management system offers the advantage of precise fuel calculations based on time, distance, winds, and fuel flow measured by other aircraft systems. When a route has been programmed into the FMS, the fuel management function is capable of displaying currently available fuel and aircraft endurance and providing an estimate of fuel remaining as the aircraft crosses each waypoint in the programmed route. A fuel management function is useful not only for making primary fuel calculations, but also for backing up calculations performed by the pilot. If there are leaks, plumbing malfunctions, or inadequate leaning, the fuel display can be deceptive. You must always land at the earliest gauge indication of low fuel in the tanks, time of normal landing, or any sign of fuel value disagreement with the flight planning. Errors can be determined when the aircraft is safely on the ground.

Initial Fuel Estimate

Many fuel management functions lack a fuel quantity sensor. Without access to this raw data of fuel quantity, fuel management functions perform calculations using an initial fuel estimate that was provided by the pilot prior to departure. *Figure 5-20* illustrates how an initial fuel estimate is given for one manufacturer's fuel management unit.

Figure 5-20. *Making an initial fuel estimate.*

It is important to make accurate estimates of initial fuel because the fuel management function uses this estimate in making predictions about fuel levels at future times during the flight. For example, if you overestimate the initial fuel by eight gallons and plan to land with seven gallons of reserve fuel, you could observe normal fuel indications from the fuel management system, yet experience fuel exhaustion before the end of the flight. The accuracy of the fuel calculations made by the fuel management function is only as good as the accuracy of the initial fuel estimate.

You must know the capacity of the aircraft fuel tanks and amount of fuel required to fill the tanks to any measured intermediate capacity (e.g., "tabs"). When full fuel capacity is entered into the fuel management system, the tanks must be filled to the filler caps. For some aircraft, even a fraction of an inch of space between the filler cap and the fuel can mean that the tanks have been filled only to several gallons under maximum capacity. Objects can plug lines, preventing the fuel from flowing to the pickup point. Some aircraft have bladders and dividers in the fuel system. A bladder can move within the tank area and not actually hold the quantity of fuel specified. Always check to ensure that the fuel servicing total matches the quantity needed to fill the tank(s) to the specified level.

Estimating Amount of Fuel on Board

Since the fuel management function's predictions are often based on the initial quantity entered, it is important to monitor the fuel gauges to ensure agreement with the fuel management function of the FMS as the flight progresses. It is always prudent to use the most conservative of these measures when estimating fuel on board.

Predicting Fuel at a Later Point in the Flight

A primary function of the fuel management function or system is to allow you to predict fuel remaining at a future time in the flight. The fuel management system uses a combination of the currently available fuel and the current rate of fuel consumption to arrive at the measures. Some units require the current or estimated fuel burn rate to be entered. Some units have optional sensors for fuel flow and/or quantity. Be absolutely sure of what equipment is installed in your specific aircraft and how to use it. Since the rate of fuel consumption instantly changes when power or mixture is adjusted, (usually with altitude) the fuel management function or system should continually update its predictions. It is common for the fuel management system to calculate fuel remaining at the arrival of the active waypoint, and the last waypoint in the route programmed into the FMS/RNAV. These measures are shown on the MFD in *Figure 5-21*.

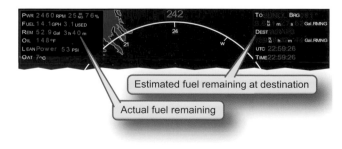

Figure 5-21. *Fuel remaining and endurance shown on an MFD.*

When no route is programmed into the FMS/RNAV, the fuel management function may not display information due to lack of data.

Determining Endurance

Most fuel management function or systems display the amount of fuel remaining, as well as the endurance of the aircraft given the current fuel flow. Most systems display the aircraft endurance in hours and minutes, as shown in *Figure 5-21*.

Some units show a fuel range ring on the MFD that indicates the distance the aircraft can fly given current fuel and fuel flow. This feature, illustrated in *Figure 5-22*, is useful for making fuel stop or alternate airport planning decisions. It may or may not include allowances for winds. Many units allow you to specify personal minimum fuel reserves. In this case, the fuel range ring indicates the point at which the aircraft will reach reserve fuel minimums.

Risk: Stretching Fuel Reserves

The availability of predictive information about fuel burn and fuel availability introduces the possibility of flying closer and

closer to fuel minimums or stretching fuel holdings farther than would be appropriate with "back of the envelope" calculations in a traditional aircraft. You must be aware of this tendency and discipline yourself to using fuel management systems to increase safety rather than stretch the limits. Refueling the aircraft offers a good opportunity to compare the amount of fuel burned with that predicted by the fuel management system and your own calculations. It is always a good exercise to determine why the fuel management function or system's numbers differ from what is actually pumped into the tank(s). Was it improper leaning? More/less winds? Are the EGT/CHT gauges indicating properly?

Other Cockpit Information System Features

Electronic Checklists

Some systems are capable of presenting checklists that appear in the aircraft operating manual on the MFD. The MFD in *Figure 5-23* depicts a pretaxi checklist while the aircraft is parked on the ramp.

In some cases, checklists presented on an MFD are approved for use as primary aircraft checklists. It is important to note

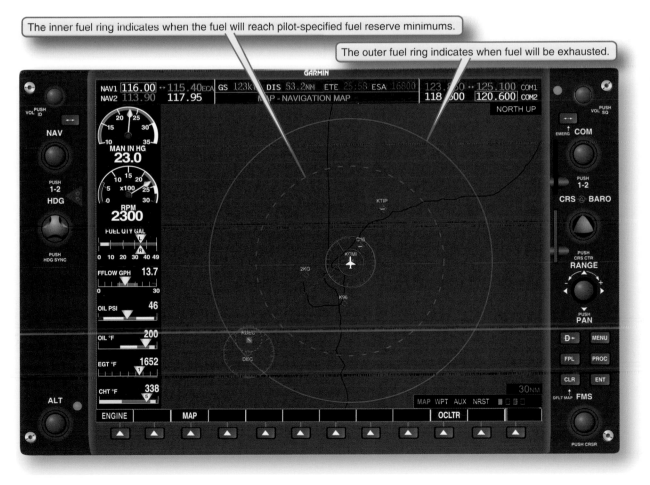

Figure 5-22. *A fuel range ring.*

Figure 5-23. *A pretakeoff checklist shown on an MFD.*

that electronic checklists are only available when the aircraft's electrical system is powered up. In almost all instances, the aircraft must have emergency checklists in paper (or plastic) form in the event of power or electrical failure. You should be well versed in the use and contents of the checklists and be able to find them in times of stress. Some climates dictate minimum battery use until the engine is started. For those circumstances, it is important to be competent in the use of paper checklists for normal procedures until the electronic checklist in the MFD becomes available.

Electronic Charts

Some systems are capable of presenting terminal and approach procedure charts on the MFD. *Figure 5-24* shows an instrument procedure presented on an MFD.

Note that the position of the aircraft is superimposed on the instrument approach chart. Electronic charts are also useful when taxiing, as they can help improve navigation on the airport surface and reduce runway and taxiway incursions.

Figure 5-24. *An instrument approach procedure shown on an MFD.*

FMS/RNAV Pages on the MFD

Some advanced avionics systems are able to draw information from the FMS/RNAV and present it on the MFD, in a larger format. The MFD in *Figure 5-25* lists nearest airports and allows you to select any airport to see information about that airport.

Figure 5-25. *A nearest airport page shown on an MFD.*

Some advanced avionics systems can integrate information from several systems in a single display. The display in *Figure 5-26* presents the route programmed into the FMS, together with fuel predictions for each waypoint made by the fuel management function.

Chapter Summary

In addition to primary flight instruments, advanced avionics (utilizing microchips) can also display landmarks, weather in real time or near real time, traffic, terrain, systems status (including fuel state) and endurance in, with, or next to the selected navigation route. None of these functions allows you to disregard the requirement "to see and avoid" traffic, obstructions, or hazardous weather. These options are designed to enhance safety—not to extend the limits of operations.

In all cases, you must accurately determine what equipment is installed and authorized. You must know the limitations of the data presented and all of the details of the displays, especially refresh rates and delays from data acquisition to presentation. The onboard data is never an adequate substitute for a timely and thorough preflight briefing.

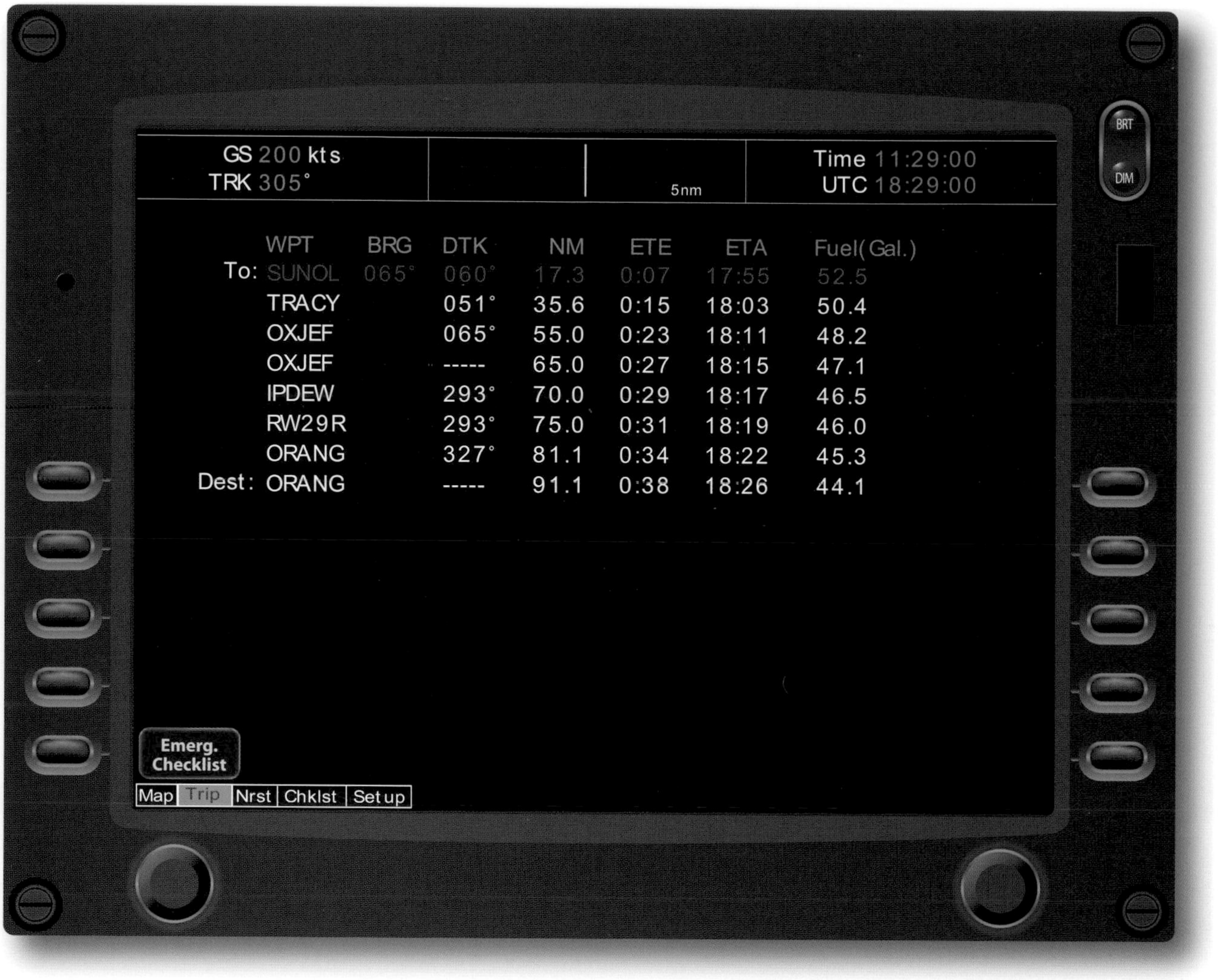

Figure 5-26. *A trip page summarizes your route and predicted fuel consumption.*

Essential Skills Checklist

Chapter 2: Electronic Flight Instruments

Primary Flight Display

1. Correctly interpret flight and navigation instrument information displayed on the PFD.

2. Determine what "fail down" modes are installed and available. Recognize and compensate appropriately for failures of the PFD and supporting instrument systems.

3. Accurately determine system options installed and actions necessary for functions, data entry and retrieval.

4. Know how to select essential presentation modes, flight modes, communication and navigation modes, and methods of mode selection and cancellation.

5. Be able to determine extent of failures and reliable information remaining available, including procedures for restoring function(s) or moving displays to the MFD or other display.

Chapter 3: Navigation

Flight Planning

1. Determine if the FMS is approved for the planned flight operation.

2. Determine if your FMS can be used as a primary navigation system for alternate requirements.

3. Understand how entries are made and canceled.

4. Understand how each unit is installed, and how it is programmed or jumpered for optional functions.

5. Determine which navigation sources are installed and functional.

6. Determine the status of the databases.

7. Program the FMS/RNAV with a flight plan, including en route waypoints, user waypoints, and published instrument procedures.

8. Review the programmed flight route to ensure it is free from error.

9. Find the necessary pages for flight information in the databases.

10. Determine which sources drive which displays or instruments, and where the selection controls are located.

11. Determine and understand how to use and program optional functions and equipment installed with FMS/RNAV basic unit.

En Route

1. Select and monitor the en route portion of the programmed flight route, determining waypoint arrival, approving turn anticipation, and waypoint sequencing.

2. Approve or select the correct course automatically displayed or manually tuned.

3. Determine if the FMS makes fuel calculations and what sensors and data entries are required to be made by the pilot.

4. Ensure that the track flown is that cleared by air traffic control (ATC).

5. Determine that the display CDI sensitivity is satisfactory for the segment being flown.

En Route Modifications

1. Proceed directly to a waypoint in the programmed route.

2. Cancel a programmed or selected waypoint or fix.

3. Select a different instrument procedure or transition.

4. Restart an approach sequence.

5. Immediately find the nearest airport or facility.

6. Edit a flight plan.

7. Enter a user waypoint.

Descent

1. Determine the descent airspeed to be used with concern to turbulence, aircraft descent profile, and powerplant cooling restrictions.

2. Program, observe, and monitor the top of descent, descent rate, and level-off altitude.

3. Plan and fly a descent to a crossing restriction.

4. Recognize and correct deviations from a planned descent path, and determine which factor changed.

Intercept and Track Course

1. Program and select a different course to the active waypoint.

2. Select the nonsequencing waypoint function (OBS, Hold, or Suspend) to select a specified navigation point.

3. Reactivate the sequencing function for route navigation.

Holds

1. Select a preprogrammed holding pattern, or nonsequencing mode.

2. Select and set up a non-preprogrammed holding pattern inbound course.

3. Determine the proper sequence of software commands for the holding pattern, transition to approach, approach, and MAP navigation.

Arcs

1. Select an approach procedure with an arc.

2. Select the course, or determine that automatic course CDI setting will occur.

GPS and RNAV (GPS) Approaches

1. Load and activate a vectored GPS or RNAV (GPS) approach.

2. Select a vectored initial approach segment.

3. Determine the correct approach minimums and identify all pertinent mode transitions.

4. Determine the published missed approach point (MAP), courses, altitudes, and waypoints to fly.

5. Determine how missed approach guidance is selected.

Course Reversals

1. Select a type of course reversal procedure.

2. Determine the correct sequence of mode control actions to be accomplished by the pilot.

Missed Approaches

1. Acknowledge a missed approach procedure.

2. Set the FMS/GPS for a return to the same approach to fly it again.

3. Select a different approach while holding at a missed approach holding waypoint.

4. Program an ATC specified hold point (user waypoint) for selection after the published MAP/hold procedure.

Ground-Based Radio Navigation

1. Select any type of ground-based radio navigation approach.

2. Correctly tune and set up the conventional navigation receiver for the procedure in number 1.

3. Correctly monitor the navaid for proper identification and validity.

4. Correctly select and be able to use the desired navigation source for the autopilot.

Chapter 4: Automated Flight Control

Climbs and Descents

1. Use the FD/autopilot to climb or descend to and automatically capture an assigned altitude.

2. Determine the indications of the ARM or capture modes, and what pilot actions will cancel those modes.

3. Determine if the system allows resetting of the armed or capture modes or if manual control is the only option after cancellation of these modes.

4. Determine the available methods of activating the altitude armed or capture mode(s).

5. Determine the average power necessary for normal climbs and descents. Practice changing the power to these settings in coordination with making the FD/autopilot mode changes.

6. Determine and record maximum climb vertical speeds and power settings for temperatures and altitudes. Ensure the values are in agreement with values in the AFM/POH for the conditions present. Make note of the highest practical pitch attitude values, conditions, and loading. Remember powerplant factors (e.g., minimum powerplant temperature, bleed air requirements) and airframe limitations (e.g., V_A in setting power).

Course Intercepts

1. Use the FD/autopilot to fly an assigned heading to capture and track a VOR and/or RNAV course.

2. Determine if the FD/autopilot uses preprogrammed intercepts or set headings for navigation course interceptions.

3. Determine the indications of navigation mode armed conditions.

4. Determine parameters of preprogrammed intercept modes, if applicable.

5. Determine minimum and maximum intercept angle limitations, if any.

Coupled Approaches

1. Use the FD/autopilot to couple to a precision approach.

2. Use the FD/autopilot to couple to a nonprecision approach.

3. Use the FD/autopilot to couple to an RNAV approach.

4. Determine the power setting required to fly the approaches.

5. Determine the power settings necessary for level-off during nonprecision approaches and go-around power settings for both precision and nonprecision approaches.

6. Determine the speeds available for the minimum recommended powerplant settings (useful for determining if an ATC clearance can be accepted for climbs, altitudes, and descents).

Miscellaneous Autopilot Topics

1. Demonstrate the proper preflight and ground check of the FD/autopilot system.

2. Demonstrate all methods used to disengage and disconnect an autopilot.

3. Demonstrate how to select the different modes and explain what each mode is designed to do and when it will become active.

4. Explain the flight director (FD) indications and autopilot annunciators, and how the dimming function is controlled.

Chapter 5: Cockpit Information Systems

Multi-Function Display

1. Program the multi-function display to show data provided by any aircraft system.

2. Determine how many data displays can be combined in one display.

3. Know how to select the PFD displays on the MFD, if available.

4. Determine which data displays can be overlaid onto the PFD and the MFD.

Glossary

Active waypoint. The waypoint being used by the FMS/RNAV as the reference navigation point for course guidance.

ADAHRS. See air data attitude and heading reference system.

ADC. See air data computer.

ADS-B. See automatic dependent surveillance—broadcast.

Advanced avionics information system. Any cockpit electronic (avionics) system designed to provide information or data to the pilot about aircraft status or position, planned routing, surrounding terrain, traffic, weather, fuel, etc. Advanced avionics systems are generally evidenced by visual displays of integrated information in lieu of mechanical or stand alone instruments for one or two data sets each.

AHRS. See Attitude Heading Reference System.

Air Data Attitude and Heading Reference System (ADAHRS). An integrated flight instrument system that combines the functions of an air data computer (the "AD" short for ADC) and an Attitude Heading Reference System (AHRS) into one unit.

Air Data Computer (ADC). The system that receives ram air, static air, and temperature information from sensors, and provides information such as altitude, indicated airspeed, vertical speed, and wind direction and velocity to other cockpit systems.

Altitude alerting system. The system that allows the pilot to receive a visual and/or auditory alert when the airplane approaches or deviates from a preselected altitude.

Altitude capture. An autopilot function that enables the autopilot to level the airplane at a selected altitude automatically.

Altitude function. An autopilot function that maintains the present altitude of the airplane.

Annunciator panel. Grouping of annunciator lights that is usually accompanied with a test switch, which when pressed illuminates all the lights to confirm they are in working order.

Approach mode/function. An autopilot function or mode that allows the pilot to capture and track any VOR radial or localizer with a higher degree of accuracy.

Area navigation (RNAV). A method of navigation that permits operations along any desired flightpath within the area of coverage of station-referenced navigation aids (e.g., GPS, VOR/DME, DME/DME, eLORAN), or within the limits of a self-contained navigation system (INS, doppler radar), or any combination.

Armed. A system mode or function that is set to become actively engaged at a later time, when certain conditions are met.

Attenuation. See radar attenuation.

Attitude Heading Reference System (AHRS). An integrated flight instrument system that provides attitude, heading, rate of turn, and slip/skid information.

Automatic dependent surveillance—broadcast (ADS-B). A surveillance system in which an aircraft or vehicle to be detected is fitted with cooperative equipment in the form of a data link transmitter. The aircraft or vehicle periodically broadcasts its GPS-derived position and other information (e.g., velocity) over the data link, which is received by a ground-based transmitter/ receiver (transceiver) for processing and display at an air traffic control facility.

Automatic mode change. Any change in mode or system status initiated by the system, rather than by a deliberate mode change action taken by the pilot.

Autopilot. An aircraft flight control system that automatically manipulates the roll, pitch, and, in some cases, the yaw control surfaces of the airplane to capture and track the route programmed into the FMS/RNAV, or altitudes, vertical speeds, headings, and courses selected by the pilot.

Autopilot flight mode annunciator. A display that presents the names of autopilot functions that are either armed or engaged. It is the only reliable source of information about what autopilot functions are in use.

Autothrottle system. A system that automatically manipulates the thrust setting of the airplane to help follow the vertical trajectory portion or selected airspeed of the planned flight route.

Autotrim system. The system that automatically adjusts the pitch trim of the airplane in response to trim commands generated by the autopilot.

Bottom-of-descent point. The end point of the descent, as calculated by the FMS/RNAV.

Broadcast weather service. A weather service that prepares weather products and transmits them to participating aircraft, also known as a data link weather service.

Chapter. Associated group of electronic "pages" of information from databases found in FMS and GPS RNAV units similar in contents, such as airports, VORs, software/unit settings, and feature selections.

Command bars. A flight director display that presents roll and pitch instructions (generally, V-shaped visual cues) to help the pilot maintain the flightpath/flight track to the selected point. The pilot keeps the airplane symbol aligned with the command bars on the flight director, or centered on the FD crossbars (e.g., older Cessna units).

Crossing restriction. A directive issued or published by air traffic control that instructs the pilot to cross a given waypoint at a specified altitude, and sometimes at a specified airspeed.

Cursor mode. The function offered by the FMS/RNAV that allows data entry into an avionics unit such as the FMS and RNAV.

Data link weather service. See broadcast weather service.

Deceleration segment. A planned portion of a descent designed to permit the aircraft to slow to meet a terminal area speed restriction, crossing restriction, or other speed restriction.

Desired track. The great circle course computed by the FMS/RNAV, it goes from the past waypoint to the next (active) waypoint.

Distance measuring equipment (DME). Line-of-sight limited airborne equipment (transceiver) using paired pulse replies from ground-based transponder to determine slant range distance by time between airborne transmission of pulses and return of pulses from the ground transponder.

DME. See distance measuring equipment.

EHSI. See electronic horizontal situation indicator.

Electronic flight instruments. Flight instruments that use electronic devices to prepare and/or present information such as airspeed, attitude, altitude, and position.

Electronic horizontal situation indicator (EHSI). Electronically generated HSI display, either CRT or LCD type, indicating all standard HSI functions on a video screen instead of using mechanical components.

eLORAN. See long range navigation.

Engaged. A system mode or function that is actively performing its function.

Error-evident display. Any display that presents information in a way that makes errors more obvious and detectable.

FD. See flight director.

Faildown. The substitute display or backup instrument mode available if the primary component fails. In some systems, for example, the MFD can substitute for the PFD if the PFD fails. The PFD information "fails down" to the MFD. In other systems, the substitute for the PFD might be the conventional standby instruments and the standby or secondary navigation CDI.

Flight director. Electronic flight calculator that analyzes the navigation selections, signals, and aircraft parameters. It presents steering instructions on the flight display as command bars or crossbars for the pilot to position the nose of the aircraft over or follow.

Flight management system (FMS). A computer system containing a database to allow programming of routes, approaches, and departures that can supply navigation data to the flight director/autopilot from various sources, and can calculate flight data such as fuel consumption, time remaining, possible range, and other values.

Fly-by waypoint. A waypoint designed to permit early turns, thus allowing the aircraft to roll out onto the center of the desired track to the next waypoint.

Fly-over waypoint. A waypoint that precludes any turn until the waypoint is overflown, and is followed by either an intercept maneuver of the next flight segment or direct flight to the next waypoint.

FMS. See flight management system.

Fuel management system or function. An advanced avionics system that assists the pilot in managing fuel by considering fuel flow, airspeed, and winds to help predict fuel remaining at each waypoint along the programmed route, total endurance, and the viability of alternative routings or diversions. Stand-alone systems may integrate the output data into the FMS/RNAV or provide a discreet display, while the fuel management function is an integral portion of the FMS/RNAV system. In either instance, the fuel data management goals are similar.

Fuel range ring. A graphical depiction of the point at which an aircraft is predicted to exhaust its fuel reserves or reach a point at which only reserve fuel remains.

Glideslope (GS) function. The autopilot function that manipulates the pitch of the aircraft to track a glideslope signal or APV guidance during a precision approach.

Global Positioning System (GPS). A Global Navigation Satellite System (GNSS) navigation system that can determine position and track the movement of an aircraft. A global positioning system (GPS) receiver must be installed on board the aircraft to receive and interpret signals from the satellite-based system.

Global Positioning System Steering (GPSS). The autopilot function that receives signals directly from the GPS/FMS/RNAV to steer the aircraft along the desired track to the active waypoint set in the GPS receiver.

GPS. See Global Positioning System.

GPS overlay approach. A conventional nonprecision approach procedure that can be flown using RNAV equipment.

GPS stand-alone approach. A nonprecision approach procedure based solely on the use of the global positioning system and an IFR-certified FMS/RNAV unit using GPS signals.

GPSS. See Global Positioning System Steering.

Great circle route. The shortest distance between two points when traveling on the surface of the earth; defined by a geometric plane that passes through the two points and the center of the earth.

Ground weather surveillance radar system. Any ground-based facility equipped to gather information about significant weather across a wide area.

Heading function. The flight director/autopilot function that steers the aircraft along a specified magnetic heading.

Highway in the sky (HITS). A type of electronic flight instrument that superimposes a 3-dimensional portrayal of a planned lateral and vertical aircraft trajectory onto an artificial horizon display.

HITS. See highway in the sky.

ILS. See instrument landing system.

Inertial navigation system (INS). Self-contained internal navigation system using sensors to measure changes in motion of aircraft, acceleration and deceleration, airspeed, altitude, and heading to maintain current position of aircraft. Also called "position keeping" because an interruption of the system requires the pilot to initialize or enter the beginning point of aircraft position reference.

INS. See inertial navigation system.

Instrument landing system. A ground-based precision instrument approach system usually consisting of a localizer, glideslope, outer marker, middle marker, and approach lights.

LNAV. Lateral (azimuth) navigation guidance. A type of navigation associated with nonprecision approach procedures or en route navigation.

LNAV/VNAV. Lateral navigation/vertical navigation minimums provided for RNAV systems that include both lateral and vertical navigation (e.g., WAAS avionics approved for LNAV/VNAV, certified barometric VNAV with IFR approach certified GPS). Procedure minimums altitude is published as DA (decision altitude).

Localizer Performance with Vertical Guidance (LPV). Provides lateral containment areas comparable to an ILS localizer and decision heights between those of LNAV/VNAV approaches and Category I ILS approaches. Approach procedure minimums that use WAAS to provide Localizer Performance with Vertical guidance (LPV). WAAS avionics equipment is required to fly to LPV minimums, which are published as DA (decision altitude).

Long range navigation (LORAN). LOng RAnge Navigation ground-based electronic navigation system using hyperbolic lines of position determined by measuring the difference in the time of reception of synchronized pulse signals from fixed transmitters. LORAN-C and eLORAN operate in the 100–110 kHz frequency band. Enhanced LORAN (eLORAN) is planned to operate using more stable timing signals and stations from other chains for greater accuracy than the current LORAN-C system.

LPV. See localizer performance with vertical guidance.

LORAN-C. See long range navigation.

Magnetic flux valve. A type of magnetometer using coils of wire as the transmitting portion of a synchronous repeating system, conventionally used to stabilize and correct a slaved gyroscopic heading (azimuth) indicator by sensing changes in the earth's magnetic field.

Magnetometer. The device that measures the strength of the earth's magnetic field to determine aircraft heading, and similar to the flux valve in function.

MFD. See multi-function display.

Mode awareness. The pilot's ability to monitor how system settings are configured throughout the flight.

Moving map. A graphical depiction of aircraft position, route programmed into the FMS/RNAV, surrounding geographical features, and any other information about the immediate flight environment such as traffic and weather that may be available from other avionics systems.

Multi-function display (MFD). A cockpit display capable of presenting information received from a variety of advanced avionics systems.

Navigation database. The information stored in the FMS/RNAV; contains most of the time-sensitive navigational information found on en route and procedural charts.

Navigation function. An autopilot function that allows you to track the route programmed in the FMS/RNAV or navigation receiver, such as a VOR radial.

Next Generation Radar System. A network of radar stations operated by the National Weather Service used to detect precipitation and wind. These data are used to prepare weather radar products that can be supplied to the cockpit via a broadcast weather service.

NEXRAD. See Next Generation Radar System.

Nonsequencing mode. The FMS/RNAV navigation mode that does not automatically sequence between waypoints in the programmed route. The nonsequencing mode maintains the current active waypoint indefinitely, and allows the pilot to specify desired track to or from that waypoint.

No-further-input prediction. A technique to help pilots maintain awareness of how advanced avionics systems are configured, and of the likely future behavior of the aircraft. No-further-input predictions are made by considering what the aircraft will do if the pilot makes no further entries or commands.

Nuisance alert. A term used to describe a "false alarm" provided by an avionics system designed to detect surrounding hazards such as proximate traffic and terrain.

OBS mode. The name for the nonsequencing mode on some FMS/RNAV units. See nonsequencing mode.

Onboard lightning detection systems. An onboard weather detection system that senses electrical discharges that suggest the presence of thunderstorm cells.

Onboard weather radar. An onboard system capable of detecting significant masses of precipitation. The primary use of weather radar is to aid the pilot in avoiding thunderstorms and their associated hazards.

Page. Any one of a collection of information displays that can appear on the FMS/RNAV unit. Every page has a title and presents information related to a particular navigation topic (e.g., airport elevation, runways, communication frequencies). Pages are usually in divisions called "chapters," which group pages of similar information by topic (e.g., airports, approaches, VORs).

PFD. See primary flight display.

Preprogrammed holding pattern (preprogrammed hold). A hold that is published as a part of an instrument procedure (e.g., approach, missed approach) and has been loaded into the FMS/RNAV. Some FMS/RNAV units automatically enter and fly the holding procedure when it is encountered. Others must be flown around the depicted holding pattern, usually by changing the heading (bug). Some units require switching to the nonsequencing or OBS mode so the active waypoint remains set to the designated holding fix.

Preprogrammed course reversal. A course reversal (commonly called a "procedure turn") that appears as part of an instrument approach procedure that has been loaded into the FMS/RNAV. Many FMS/RNAV units automatically attempt to perform the course reversal procedure when it is encountered. Others require the pilot to navigate the depicted procedural track manually or by using the heading mode to fly the depicted track.

Primary flight display (PFD). An electronic flight display that presents the primary flight instruments, navigation instruments, and other information about the status of the flight in one integrated presentation. Primary-secondary task inversion. Situation in which the pilot ceases to monitor the situation directly and simply listens for system alerts.

RA. See resolution advisory.

Radar attenuation. The absorption or reflection of radar signals by a weather cell, preventing that radar from detecting any additional cells that might lie behind the first cell.

RAIM. See receiver autonomous integrity monitoring. Receiver autonomous integrity monitoring. The self-monitoring function performed by a TSO-129 certified GPS receivers to ensure that adequate GPS signals are being received at all times. The GPS will alert the pilot whenever the integrity monitoring determines that the GPS signals do not meet the criteria for safe navigation use.

Receiver autonomous integrity monitoring. The self-monitoring function performed by a TSO-129 certified GPS receiver to ensure that adequate GPS signals are being received at all times. The GPS alerts the pilot whenever the integrity monitoring determines that the GPS signals do not meet the criteria for safe navigation use.

Resolution advisory (RA). A warning issued by the Traffic Collision Avoidance System (TCAS) indicating an immediate threat of collision with another aircraft. This warning takes the form of a command to perform a vertical avoidance maneuver (e.g., "Climb! Climb!") These commands are products of TCAS II equipment. These commands take precedence over ATC instructions, but must be reported to ATC immediately upon receipt and execution.

Risk homeostasis. A term coined by psychologist Gerald J. S. Wilde, a tendency for humans to seek target levels of risk.

RNAV. See Area Navigation.

RNAV (GPS) approach. An approach procedure based on GPS signals for guidance.

Route discontinuity. A point of uncertainty in a route that has been programmed into an FMS/RNAV. Most systems display this message when there is no routing to connect the last waypoint to the next point, or there is a missing next point. FMS/RNAV units will not plan to go direct unless certain programming parameters are met.

Sequencing mode. The FMS/RNAV mode that automatically sequences along the waypoints in the programmed route. The sequencing mode alerts the pilot to upcoming waypoints, and offers guidance to each successive waypoint in the route.

Stand-alone approach. An instrument approach that relies solely on the use of RNAV equipment. If flown with GPS/WAAS enabled certified equipment in accordance with TSO-C145A or TSO-146A installed in accordance with the provisions of AC 20-130A or 138A, no conventional navigation equipment alternate approach (VOR/ILS) requirements are necessary, as when flying with TSO-C129 certified equipment.

Subpage. An additional page of information about a particular topic that can be displayed on an FMS/RNAV. Many pages require the use of several subpages to show all information pertaining to any one topic.

SUSP. See suspend mode.

Suspend mode. For some FMS/RNAV units, the name used to describe the nonsequencing mode when it has been automatically set by the computer, or pilot.

TA. See traffic advisory.

Tape display. A vertical display format used to portray, for example, airspeed and altitude on many primary flight displays. Also used for vertical speeds and many other value displays such as power settings and powerplant speeds.

TAWS. See terrain awareness and warning system.

TCAS. See Traffic Alert and Collision Avoidance System.

Terminal Arrival Area. The published or assigned track by which aircraft are transitioned from the en route structure to the terminal area. A terminal arrival area consists of a designated volume of airspace designed to allow aircraft to enter a protected area with obstacle clearance and signal reception guaranteed where the initial approach course is intercepted.

Terminal mode. One name used for the FMS/RNAV sensitivity mode in which the aircraft is operating within 30 NM of an airport. In terminal mode, the required navigation performance sensitivity of the course deviation indicator becomes 1 NM. Also called approach arm mode.

Terrain and obstacle database. An electronic database storing details of the significant terrain features and obstacles that could potentially pose a threat to aircraft flight. Some obstructions, especially manmade, may not be in the database, even if it is current. Do not plan a flight based on dependence on the database to keep the aircraft clear of obstacles and obstructions to navigation.

Terrain Awareness and Warning System (TAWS). An onboard system that can alert the pilot to a number of potential hazards presented by proximate terrain such as excessive rate of descent, excessive closure rate to terrain, and altitude loss after takeoff.

Terrain display. A pictorial display that shows surrounding terrain and obstacles that present a potential threat to your aircraft, given your present altitude. Draws terrain information from a terrain and obstacle database.

Terrain inhibit switch. A switch that allows the pilot to suppress all visual and auditory warnings given by a terrain system. Often used to silence nuisance alerts when in deliberate operation in the vicinity of terrain.

Terrain system. Any cockpit system that provides the pilot with a pictorial view of surrounding terrain, and in some cases, visual and/or auditory alerts when the aircraft is operating in close proximity to terrain.

TIS. See traffic information service.

Top-of-descent point. The point that the RNAV computer calculates to be the ideal location at which to begin a descent to the planned crossing restriction, given the descent speed and rate that has been entered by the pilot.

Topographical database. A volume of information stored in an advanced cockpit system that details the topographical features of the earth's surface. Used by several systems to assess aircraft position and altitude with respect to surrounding terrain.

Traffic Advisory (TA). A warning issued by a traffic system that alerts the pilot to other aircraft that have moved within a prescribed "safety zone" that surrounds the aircraft.

Traffic Alert and Collision Avoidance System. An onboard system that detects the presence of some aircraft operating in the vicinity of the airplane by querying the transponders of nearby aircraft and presenting their locations and relative altitudes on a display. Alerts and warnings are issued when nearby aircraft are deemed to be a threat to safety. Traffic advisory systems such as ADS-B are an offshoot of newer technologies, but do not yet offer the reliability or accuracy of proven, certified TCAS units.

Traffic data system. An advanced avionics system designed to aid the pilot in visually acquiring and maintaining awareness of nearby aircraft that pose potential collision threats.

Traffic display. A pictorial display showing any aircraft operating in the vicinity that have been detected by a traffic data system.

Traffic Information Service (TIS). A groundbased advanced avionics traffic display system which receives transmissions on locations of nearby aircraft from radar-equipped air traffic control facilities and provides alerts and warnings to the pilot.

Turn anticipation. The function performed by FMS/RNAV units to advise the pilot when to begin a turn to the next waypoint in the programmed flight route to avoid overshooting the programmed track.

Vectors to final. A function of FMS/RNAV units allowing the pilot to perform a vectored approach procedure without being required to switch to the nonsequencing mode manually and set the active waypoint and course.

Vertical speed mode. An FD/autopilot mode that allows constant-rate climbs and descents by selecting a vertical speed on the flight director or autopilot control panel.

Very high frequency omnidirectional range (VOR). Ground-based electronic navaid transmitting 360° azimuth signals on assigned frequencies ranging from 108.0 to 117.9 mHz to serve as the basis for the National Airspace System (NAS). The signals are identified by discreet Morse code identifiers and may have voice capabilities for ATC and FSS/AFSS communications.

VOR. See very high frequency omnidirectional range.

WAAS. See Wide Area Augmentation System.

Waypoint. A named geographical location used to define routes and terminal area procedures. Modern advanced navigation avionics such as FMS/RNAV units are able to locate and follow courses to and from waypoints that occur anywhere in the airspace.

Waypoint alerting. The function performed by the FMS/RNAV to alert the pilot at some time or distance prior to, or when reaching, the active waypoint.

Waypoint sequencing. The action performed by the FMS/RNAV when the aircraft effectively has reached the active waypoint, and then automatically switches to the next waypoint in the programmed route. (See turn anticipation.)

Wide Area Augmentation System (WAAS). A ground and satellite integrated navigational error correction system that provides accuracy enhancements to signals received from the Global Positioning System. WAAS provides extremely accurate lateral and vertical navigation signals to aircraft equipped with GPS/WAAS-enabled certified equipment in accordance with TSO-C145A or TSO-146A installed in accordance with the provisions of Advisory Circular (AC) 20-130A or AC 20-138A.

ALSO AVAILABLE

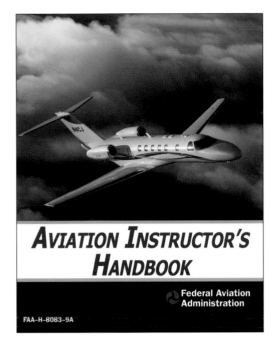

Aviation Instructor's Handbook
FAA-H-8083-9A
Federal Aviation Administration
The official FAA guide—an essential reference for all instructors.
$14.95 Paperback • 228 pages

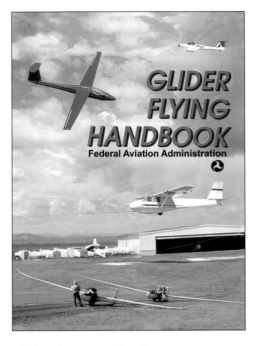

Glider Flying Handbook
Federal Aviation Administration
For certified glider pilots and students attempting certification in the glider category, this is an unparalleled resource.
$24.95 Paperback • 240 pages

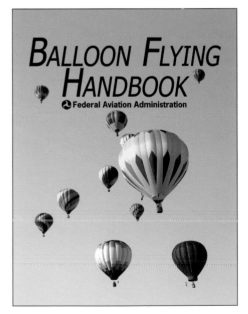

Balloon Flying Handbook
FAA-H-8083-11A
Federal Aviation Administration
Essential knowledge necessary for safe piloting at all experience levels. Includes useful illustrations, graphs, and charts.
$16.95 Paperback • 256 pages

Plane Sense
A Beginner's Guide to Owning and Operating Private Aircraft, FAA-H-8083-19A
Federal Aviation Administration
The definitive guide to buying, owning, and maintaining your private aircraft.
$12.95 Paperback • 112 pages

ALSO AVAILABLE

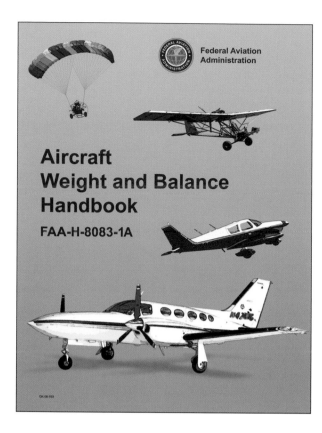

Aircraft Weight and Balance Handbook
FAA-H-8083-1A

Federal Aviation Administration

The official FAA guide to aircraft weight and balance.

Aircraft Weight and Balance Handbook is the official U.S. government guidebook for pilots, flight crews, and airplane mechanics. Beginning with the basic principles of aircraft weight and balance control, this manual goes on to cover the procedures for weighing aircraft in exacting detail. It also offers a thorough discussion of the methods used to determine the location of an aircraft's empty weight and center of gravity (CG), including information for an A&P mechanic to determine weight changes caused by repairs or alterations.

With instructions for conducting adverse-loaded CG checks and for determining the amount and location of ballast needed to bring CG within allowable limits, the *Aircraft Weight and Balance Handbook* is essential for anyone who wishes to safely weigh and fly aircraft of all kinds.

$9.95 Paperback • 96 pages

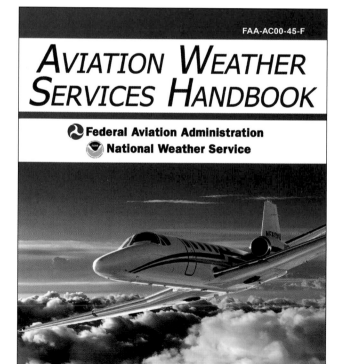

Aviation Weather Services Handbook
FAA-AC00-45-F

Federal Aviation Administration and
National Weather Service

A necessary tool for aviators of all skill levels and professions. Includes useful photographs, diagrams, charts, and illustrations.

This official handbook provides an authoritative tool for pilots, flight instructors, and those studying for pilot certification. From both the Federal Aviation Administration and the National Weather Service, this newest edition offers up-to-date information on the interpretation and application of advisories, coded weather reports, forecasts, observed and prognostic weather charts, and radar and satellite imagery. Expanded to 400 pages, this edition features over 200 color and black-and-white photographs, satellite images, diagrams, charts, and other illustrations. With extensive appendixes, forecast charts, aviation website recommendations, and supplementary product information, this book is an exhaustive resource no aviator or aeronautical buff should be without.

$19.95 Paperback • 218 pages

ALSO AVAILABLE

Seaplane, Skiplane and Float/Ski Equipped Helicopter Operations Handbook
FAA-H-8083-23

Federal Aviation Administration

The ultimate guide to water-related aircraft piloting.

This comprehensive handbook provides the most up-to-date, definitive information on piloting water-related aircraft. Along with full-color photographs and illustrations, detailed descriptions make complicated tasks easy to understand while the index and glossary provide the perfect references for finding any topic and solving any issue.

The FAA leaves no question unanswered in the most complete book on how to fly water-related aircraft available on the market. The *Seaplane, Skiplane, and Float/Ski Equipped Helicopter Operations Handbook* is the perfect addition to the bookshelf of all aircraft enthusiasts, FAA fans, and novice and experienced pilots alike.

$12.95 Paperback • 96 pages

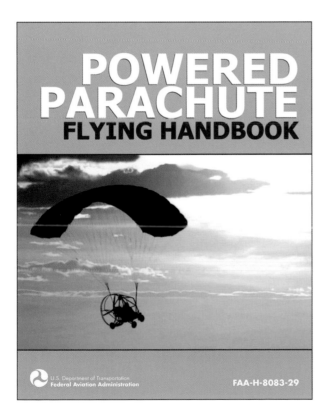

Powered Parachute Flying Handbook
FAA-H-8083-29

Federal Aviation Administration

From the FAA, the only handbook you need to learn to fly a powered parachute.

As far back as the twelfth century, people have loved to parachute. From China's umbrella and Leonardo da Vinci's pyramid-shaped flying device to the first airplane jump in 1912, the urge to leap and soar with the wind has long been a part of history. Parachuting has come a long way since its earliest days due to technological advances, and now more people than ever are taking up this incredible sport. With the *Powered Parachute Flying Handbook* you can make your flying ambitions a reality.

$24.95 Paperback • 160 pages

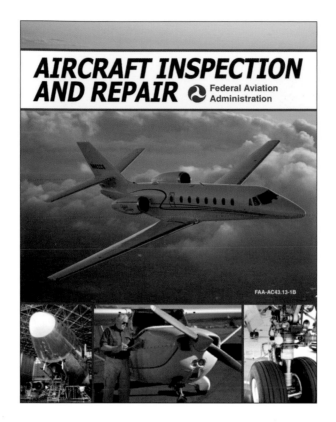

Aircraft Inspection and Repair
Acceptable Methods, Techniques, and Practices
FAA-AC-213-13-1B

Federal Aviation Administration

With every deadly airplane disaster or near-miss, it becomes more and more clear that proper inspection and repair of all aircrafts is essential to safety in the air. When no manufacturer repair or maintenance instructions are available, the Federal Aviation Administration deems *Aircraft Inspection and Repair* the one–stop guide to all elements of maintenance: preventive, rebuilding, and alteration. With detailed information on structural inspection, protection, and repair—including aircraft systems, hardware, fuel, engines, and electrical systems—this comprehensive guide is designed to leave no vital question on inspection and repair unanswered. Sections include:

- Wood, fabric, plastic, and metal structures
- Testing of metals and repair procedures
- Welding and brazing, including fire explosion and safety
- Nondestructive inspection (NDI)
- Application of magnetic particles
- Common corrosive elements and corrosion proofing
- Aircraft hardware, from nuts and bolts to washers and pins
- Engines, fuel, exhaust, and propellers
- Aircraft systems and components
- Electrical systems

$24.95 Paperback • 768 pages

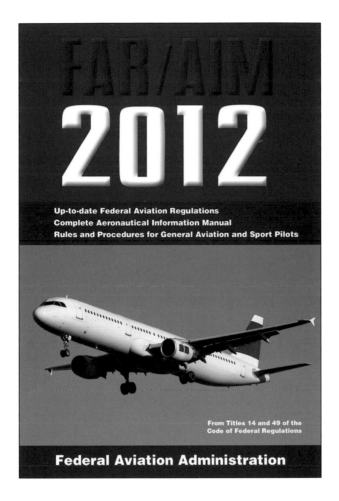

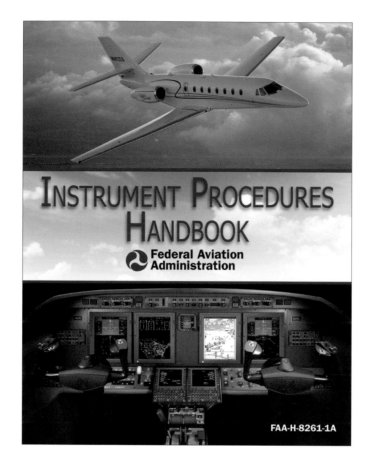

Instrument Procedures Handbook

FAA-H-8261-1A

Federal Aviation Administration

Designed as a technical reference for instrument-rated pilots who want to maximize their skills in an "Instrument Flight Rules" environment, the Federal Aviation Administration's *Instrument Procedures Handbook* contains the most current information on FAA regulations, the latest changes to procedures, and guidance on how to operate safely within the National Airspace System in all conditions. In-depth sections cover takeoffs and departures, en route operations, arrivals and approach, system improvement plans, and helicopter instrument procedures. Thorough safety information covers relevant subjects such as runway incursion, land and hold short operations, controlled flight into terrain, and human factors. Featuring an index, an appendix, a glossary, full-color photos, and illustrations, the *Instrument Procedures Handbook* is a valuable training aid and reference for pilots, instructors, and flight students, and the most authoritative book on instrument use anywhere.

$19.95 Paperback • 296 pages

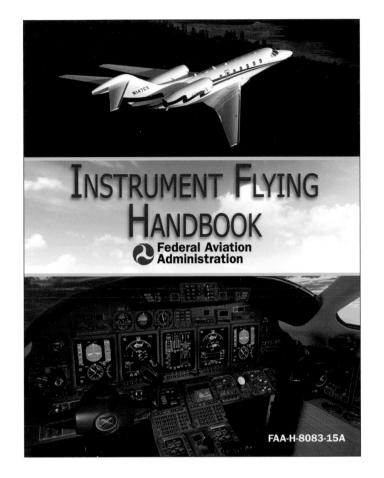

Instrument Flying Handbook

FAA-H-8083-15A

Federal Aviation Administration

The Federal Aviation Administration's *Instrument Flying Handbook* provides pilots, student pilots, aviation instructors, and controllers with the knowledge and skills required to operate in instrument meteorological conditions. This up–to–date edition is illustrated with full–color graphics and photographs, and covers topics such as basic atmospheric science, the air traffic control system, spatial disorientation, and optical illusions, flight support systems, and emergency responses. Since many questions on FAA exams are taken directly from the information presented in this text, the *Instrument Flying Handbook* is a great study guide for potential pilots looking for certification, and the perfect gift for any aircraft or aeronautical buff.

$19.95 Paperback • 392 pages

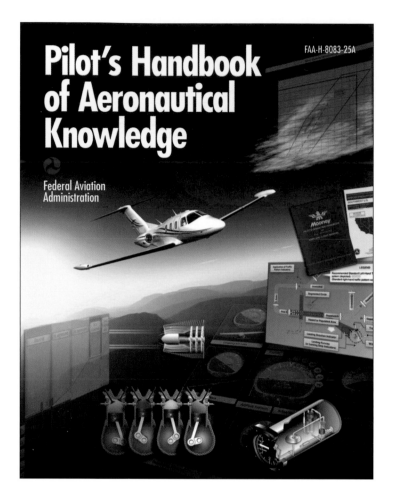